Electronic Processes in Solids

TECHNOLOGY PRESS RESEARCH MONOGRAPHS

ELECTRONIC PROCESSES IN SOLIDS
By Pierre R. Aigrain

THE DYNAMIC BEHAVIOR OF THERMOELECTRIC DEVICES
By Paul E. Gray

HYDROMAGNETIC CHANNEL FLOWS
By Lawson P. Harris

PROCESSING NEUROELECTRIC DATA
By Communications Biophysics Group of Research Laboratory of Electronics and William M. Siebert

MATHEMATICAL PROGRAMMING AND ELECTRICAL NETWORKS
By Jack B. Dennis

CIRCUIT THEORY OF LINEAR NOISY NETWORKS
By Hermann A. Haus and Richard B. Adler

NONLINEAR PROBLEMS IN RANDOM THEORY
By Norbert Wiener

Prepared by
ROLAND J. COELHO
GIANNI ASCARELLI

Electronic Processes in Solids

Based on lectures given by
PIERRE R. AIGRAIN
Visiting Webster Professor of
Electrical Engineering
Massachusetts Institute of Technology
May, 1957

The M.I.T. Press
Cambridge, Massachusetts

Library of Congress Catalog Card Number: 60-11713

Printed in the United States of America

Foreword

There has long been a need in science and engineering for systematic publication of research studies larger in scope than a journal article but less ambitious than a finished book. Much valuable work of this kind is now published only in a semiprivate way, perhaps as a laboratory report, and so may not find its proper place in the literature of the field. The present contribution is the seventh of the Technology Press Research Monographs, which we hope will make selected timely and important research studies accessible to libraries and to the independent worker.

J. A. Stratton

Preface

This book is based on a series of eight lectures given in May, 1957, by Professor Pierre R. Aigrain, while he served as Visiting Webster Professor in Electrical Engineering at the Massachusetts Institute of Technology.

Although Professor Aigrain's lectures form the very basis of this book, we have taken the liberty of extending some of the topics, and presenting others in a slightly different way. The reason for these alterations is the desire to make the material understandable to any reader possessing a fair background in calculus and elementary wave mechanics. Since Professor Aigrain had to cover a broad field of solid state physics during a few lectures, he had to assume that the audience was familiar with the necessary background. In our efforts to reach a wider group, however, we are probably responsible for some mistakes, for which we apologize.

Some of the material of these lectures is already contained in the monograph *Les Semiconducteurs* by P. Aigrain and F. Englert, Monographies Dunod, Paris, 1958. We also acknowledge the use of Technical Reports IV, *The Energy Band Method*, and V, *The Perturbed Periodic Lattice*, by J. C. Slater (M.I.T. Solid State and Molecular Theory Group, 1953) and *Introduction to Solid State Physics* by C. Kittel (second edition, John Wiley and Sons, New York, 1956).

Roland Coelho
Gianni Ascarelli

Cambridge, Massachusetts
August, 1959

Contents

1. Introduction—The Crystal Hamiltonian

The main objective of this book is to present the physical background which is needed for the study of electric conduction phenomena in crystals.

A crystal is made of a large number of atomic cores which can vibrate around equilibrium positions forming a regular lattice. The cores are surrounded by electrons which are more-or-less tightly bound to particular nuclei.

Since our goal is to find whether and how electrons can move through the crystal under the influence of an external electric field, it is logical to study first the motion of the most loosely bound electrons in the absence of applied field, and then to see how the state of affairs is modified when a field is applied.

The speed of electrons in solids being much smaller than the speed of light, the magnetic interactions, which are proportional to v/c, are very small. Hence, the magnetic forces resulting classically from the motion of the electrons themselves as well as from an external magnetic field will be neglected in our discussion. However, the spin of the electrons will be taken into account, in the choice of the linear combination of wave functions which is antisymmetric in the interchange of any two electrons, as prescribed by the Pauli exclusion principle.

The problem of the motion of the nuclei and electrons in a system of N interacting atoms can be formulated in a concise fashion. Let $\mathbf{R}_a$, a vector in a $3N$-dimensional space, represent the co-ordinates of the N nuclei, and $\mathbf{r}_i$, a vector in a $3P$-dimensional space, represent the co-ordinates of the P electrons. The Hamiltonian operator of the system, in which no crystalline symmetry is yet introduced, can be written

$$H = \sum_a -\frac{\hbar^2}{2M_a}\nabla_a^2 + \sum_i -\frac{\hbar^2}{2m}\nabla_i^2 + \sum_{a,b}{}' V_{ab} + \sum_{i,j}{}' V_{ij} + \sum_{a,i} V_{ai} \tag{1.1}$$

where V_{ab} is the interaction potential of two nuclei, V_{ij} that of two electrons, and V_{ai} that of an electron and a nucleus.

Let us denote by $\phi_r(\mathbf{r}_1,\cdots,\mathbf{r}_i,\cdots;\mathbf{R}_1,\cdots,\mathbf{R}_a,\cdots)$ and E_r the eigenfunctions and eigenvalues of H, respectively. If we could find these eigenfunctions and eigenvalues, we should know everything about the motion of the electrons and the nuclei in the crystal. Obviously, we shall have to reduce this formidable task into a number of smaller problems. A first and fundamental step toward this goal is made possible by the Born-Oppenheimer approximation, which is discussed next.

2. The Born-Oppenheimer Approximation

If the nuclei are assumed fixed at their average positions, we only have to consider the Hamiltonian for the electrons, which is

$$H_e = \sum_i -\frac{\hbar^2}{2m}\nabla_i^2 + \sum_{i,j}{}' V_{ij} + \sum_{a,i} V_{ai} \tag{2.1}$$

Let us call $\psi_s(\mathbf{r}_1,\cdots,\mathbf{r}_i,\cdots;\mathbf{R}_1,\cdots,\mathbf{R}_a,\cdots)$ and $\mathscr{E}_s(\mathbf{R}_1,\cdots,\mathbf{R}_a,\cdots)$ the eigenfunctions and eigenvalues, respectively, of H_e. The nuclear co-ordinates $\mathbf{R}_1,\cdots,\mathbf{R}_a,\cdots$ enter as parameters.

Let us now assume that $\phi_r(\mathbf{r}_1,\cdots,\mathbf{r}_i,\cdots;\mathbf{R}_1,\cdots,\mathbf{R}_a,\cdots)$ can be written as a product thus*

$$\phi_r(\mathbf{r}_1,\cdots,\mathbf{r}_i,\cdots;\mathbf{R}_1,\cdots,\mathbf{R}_a,\cdots) \equiv g_{r,s}(\mathbf{R}_1,\cdots,\mathbf{R}_a,\cdots)\psi_s(\mathbf{r}_1,\cdots,\mathbf{r}_i,\cdots;\mathbf{R}_1,\cdots,\mathbf{R}_a,\cdots) \tag{2.2}$$

* From now on, the parentheses in relation to ϕ_r, $g_{r,s}$, ψ_s, and $\mathscr{E}_s$ will usually be omitted for clarity; however, it is important that the reader understands clearly their significance.

Using this in Eq. 2.3, we have

$$H_r\phi_r = E_r\phi_r \tag{2.3}$$

where

$$H = H_e + \sum_a -\frac{\hbar^2}{2M_a}\nabla_a{}^2 + \sum_{a,b}{}' V_{ab}$$

and remembering that

$$H_e\psi_s = \mathscr{E}_s\psi_s \tag{2.4}$$

we obtain

$$\mathscr{E}_s g_{r,s}\psi_s + \sum_{a,b}{}' V_{ab}g_{r,s}\psi_s + \sum_a -\frac{\hbar^2}{2M_a}[g_{r,s}\nabla_a{}^2\psi_s + \psi_s\nabla_a{}^2 g_{r,s} + 2(\boldsymbol{\nabla}_a g_{r,s}\cdot\boldsymbol{\nabla}_a\psi_s)] = E_r g_{r,s}\psi_s \tag{2.5}$$

If we multiply all the terms of Eq. 2.5 by $\psi_s{}^*$ and integrate over all electronic co-ordinates, remembering that the solutions of Eq. 2.4 form a complete orthonormal set, we obtain

$$\mathscr{E}_s g_{r,s} + \sum_a V_a g_{r,s} + \sum_a -\frac{\hbar^2}{2M_a}\left[g_{r,s}\int\psi_s{}^*\nabla_a{}^2\psi_s\,d^3\mathbf{r}_i + 2\nabla_a g_{r,s}\cdot\int\psi_s{}^*\,\boldsymbol{\nabla}_a\psi_s\,d^3\mathbf{r}_i\right] + \sum_a -\frac{\hbar^2}{2M_a}\nabla_a{}^2 g_{r,s} = E_r g_{r,s}$$

or, by reorganization of terms,

$$\sum_a -\frac{\hbar^2}{2M_a}\nabla_a{}^2 g_{r,s} + \mathscr{E}_s g_{r,s} + \sum_a V_a g_{r,s} = E_r g_{r,s} + \sum_a \frac{\hbar^2}{2M_a}\left[g_{r,s}\int\psi_s{}^*\,\nabla_a{}^2\psi_s\,d^3\mathbf{r}_i + 2\boldsymbol{\nabla}_a g_{r,s}\cdot\int\psi_s{}^*\,\boldsymbol{\nabla}_a\psi_s\,d^3\mathbf{r}_i\right] \tag{2.6}$$

Let us focus our attention on the last terms on the right-hand side of Eq. 2.6. The energy associated with them is

$$\sum_a \frac{\hbar^2}{2M_a}\left[\iint g_{r,s}{}^*g_{r,s}\,d^3\mathbf{R}_a\psi_s{}^*\,\nabla_a{}^2\psi_s\,d^3\mathbf{r}_i + 2\iint g_{r,s}{}^*\,\boldsymbol{\nabla}_a g_{r,s}\,d^3\mathbf{R}_a\cdot\psi_s{}^*\,\boldsymbol{\nabla}_a\psi_s\,d^3\mathbf{r}_i\right] \tag{2.6a}$$

If this quantity is small compared to E_r, the set of equations 2.2 is a good approximation for the eigenfunctions of H.

We shall now discuss the validity of this approximation under the assumption that electrons are either tightly bound to the nuclei or completely free.

In the first case, the terms of the type V_{ij} in the potential can be disregarded, and the functions ψ_s will be products of one-electron wave functions, each of which depends only on the difference $(\mathbf{R}_a - \mathbf{r}_i)$. Under these conditions

$$\nabla_a{}^n\psi_s(\mathbf{r}_i - \mathbf{R}_a) = (-1)^n \nabla_i{}^n\psi_s$$

so that the terms $\hbar^2/2M_a \int \psi_s{}^* \nabla_a{}^2\psi_s \, d^3\mathbf{r}_i$ in Eq. 2.6a take the form

$$\frac{\hbar^2}{2M_a}\int \psi_s{}^* \nabla_i{}^2\psi_s \, d^3\mathbf{r}_i = \frac{m}{M_a}\int \psi_s{}^* \frac{\hbar^2}{2m} \nabla_i{}^2\psi_s \, d^3\mathbf{r}_i = \frac{m}{M_a}\left\langle \begin{matrix}\text{kinetic energy}\\ \text{of electron}\end{matrix}\right\rangle_{\text{av}}$$

Hence, these terms contribute to the energy of a vibrating atom by m/M_a of the average kinetic energy of an electron, that is, 10^{-5} of this energy in the case of germanium.

The second term that was neglected is

$$\sum_a \frac{\hbar^2}{M_a}\iint g_{r,s}^* \, \boldsymbol{\nabla}_a g_{r,s} \, d^3\mathbf{R}_a \cdot \psi_s{}^* \, \boldsymbol{\nabla}_a\psi_s \, d^3\mathbf{r}_i$$

Here, we can replace $\boldsymbol{\nabla}_a$ by $-\boldsymbol{\nabla}_i$, so that the integral with respect to $\mathbf{r}_i$ becomes $-\int \psi_s{}^* \boldsymbol{\nabla}_i\psi_s \, d^3\mathbf{r}_i$, which is nothing but $i/\hbar$ times the average momentum of the electrons, while the integral involving $g_{r,s}$ is merely $i/\hbar$ times the average momentum of the atoms. Consequently, the second term of Eq. 2.6a can be rewritten as

$$\sum_a -\frac{1}{M_a}\left\langle \begin{matrix}\text{momentum}\\ \text{of nuclei}\end{matrix}\right\rangle_{\text{av}} \left\langle \begin{matrix}\text{momentum}\\ \text{of electrons}\end{matrix}\right\rangle_{\text{av}}$$

Under thermal equilibrium, we have

$$\left\langle \frac{p_e{}^2}{m} \right\rangle = \left\langle \frac{p_a{}^2}{M_a} \right\rangle$$

so that

$$\langle p_e \rangle \sim \sqrt{m/M_a}\langle p_a \rangle \sim 3 \times 10^{-3}\langle p_a \rangle$$

in germanium. Hence, the second term of Eq. 2.6a contributes $\sqrt{m/M_a}$ per cent to the average energy of the atoms, that is, about 0.3 per cent in the case of germanium, and its neglect is justified in the tight-binding approximation.

In the second case (free electrons), $\nabla_a{}^n\psi_s \equiv 0$ because the terms V_{aj} in H_e are zero, and $\phi_r = g_{r,s}\psi_s$ is an exact solution of the system.

The actual cases are intermediate between the cases of tightly bound and free electrons. Since we have shown that the terms which are neglected in the Born-Oppenheimer approximation are small even in the tight-binding assumption, they can be treated as perturbations capable of inducing transitions between the eigenstates of the approximate nuclear Hamiltonian:

$$H_a' = \sum - \frac{\hbar^2}{2M_a} \nabla_a{}^2 + \mathscr{E}_s + \sum_a V_a \tag{2.7}$$

The Schrödinger equation of the crystal has been effectively separated into two parts: one of which depends only on the nuclear co-ordinates; the other one, only on the electronic co-ordinates but contains the nuclear co-ordinates as parameters.

3. The Harmonic Vibrations of the Nuclei

We shall now solve the Schrödinger equation corresponding to the Hamiltonian of Eq. 2.7:

$$\left[\sum_a - \frac{\hbar^2}{2M_a} \nabla_a{}^2 + \mathscr{E}_s(\mathbf{R}_a) + \sum_a V_a\right] g_{r,s} = E_r g_{r,s} \tag{3.1}$$

It is convenient to denote the sum $\mathscr{E}_s(\mathbf{R}_a) + \sum_a V_a$ by $\mathscr{V}(\mathbf{R}_a)$, and since the lattice constant is practically invariant, $\mathscr{V}(\mathbf{R}_a)$ can be expanded in terms of the displacements $\mathbf{u}_a = \mathbf{R}_a - \mathbf{R}_{a,0}$ from the average positions $\mathbf{R}_{a,0}$.

Because of the definition of $\mathbf{R}_{a,0}$, this expansion does not contain linear terms in the $\mathbf{u}$'s, so that we can write

$$\mathscr{V}(\mathbf{R}_a) = \mathscr{V}(\mathbf{R}_{a,0}) + \sum_{a,b} \tfrac{1}{2} A_{ab} u_a u_b + \text{(higher-order terms)} \tag{3.2}$$

Here, u_a and u_b refer to the magnitude of the corresponding vectors, and their respective orientation is taken up in the coefficients A_{ab}.

By proper transformation of the co-ordinates, the quadratic terms $A_{ab} u_a u_b$ can be reduced to squares. If we assume that the terms having powers of the $\mathbf{u}$'s higher than the second are negligible, the displacements can be written by analogy with a classical harmonic motion:

$$\mathbf{u}_a = \sum_{\mathbf{k}} \boldsymbol{\xi}_{\mathbf{k}}(t) \exp(i\mathbf{k}\cdot\mathbf{R}_a) \tag{3.3}$$

where

$$\boldsymbol{\xi}_{\mathbf{k}}(t) = \xi_{\mathbf{k}} \mathbf{v}_{\mathbf{k}} \exp(i\omega_{\mathbf{k}} t)$$

$\mathbf{v_k}$ being a unit vector collinear with $\boldsymbol{\xi}_{\mathbf{k}}(0)$. The $\mathbf{u}$'s are thus described by a superposition of harmonic displacements with various frequencies. From the correspondence principle we know that the eigenfrequencies of the quantum-mechanical Hamiltonian are the frequencies of the normal modes of vibration. The quantum-mechanical Hamiltonian can thus be written in terms of the $\boldsymbol{\xi}_{\mathbf{k}}$ and $\omega_{\mathbf{k}}$, so that Eq. 3.1 becomes

$$\left[\sum_{\mathbf{k}}\left(-\frac{\hbar^2}{2M_a}\frac{\partial^2}{\partial\xi_{\mathbf{k}}^2}+\frac{M_a}{2}\,\omega_{\mathbf{k}}\xi_{\mathbf{k}}^2\right)\right]g_{r,s}(\boldsymbol{\xi}_1,\cdots,\xi_{\mathbf{k}},\cdots)$$
$$= E_r g_{r,s}(\boldsymbol{\xi}_1,\cdots,\boldsymbol{\xi}_{\mathbf{k}},\cdots) \quad (3.4)$$

This equation is clearly separable into a set of differential equations if we write

$$g_{r,s}(\boldsymbol{\xi}_1,\cdots,\boldsymbol{\xi}_{\mathbf{k}},\cdots) = \prod_{\mathbf{k}} \underset{\mathbf{k}}{g_{r,s}}(\boldsymbol{\xi}_{\mathbf{k}}) \quad (3.5)$$

The total energy is $E_r = \sum_{\mathbf{k}} \mathscr{E}_{r,\mathbf{k}}$, where each of the $\mathscr{E}_{r,\mathbf{k}}$ is one of the eigenvalues of a linear harmonic oscillator.

In order to solve the individual Schrödinger equations for the harmonic oscillator, it is both convenient and instructive to introduce the creation and destruction operators, since these operators will be used again later.

The equation corresponding to $\boldsymbol{\xi}_{\mathbf{k}}$ can be written as

$$\left[\frac{\hbar^2}{2M_a}\frac{\partial^2}{\partial\xi_{\mathbf{k}}^2}+\mathscr{E}-\tfrac{1}{2}M_a\omega_{\mathbf{k}}^2\xi_{\mathbf{k}}^2\right]\underset{\mathbf{k}}{g_{r,s}}(\boldsymbol{\xi}_{\mathbf{k}}) = 0 \quad (3.6)$$

Introducing the operators

$$a_{\mathbf{k}} = \left(\frac{\hbar^2}{2M_a}\right)^{1/2}\frac{\partial}{\partial\xi_{\mathbf{k}}}+\left(\frac{M_a\omega_{\mathbf{k}}^2}{2}\right)^{1/2}\xi_{\mathbf{k}} \quad (3.7a)$$

$$a_{\mathbf{k}}^+ = \left(\frac{\hbar^2}{2M_a}\right)^{1/2}\frac{\partial}{\partial\xi_{\mathbf{k}}}-\left(\frac{M_a\omega_{\mathbf{k}}^2}{2}\right)^{1/2}\xi_{\mathbf{k}} \quad (3.7b)$$

we have

$$a_{\mathbf{k}}^+a_{\mathbf{k}} = \frac{\hbar^2}{2M_a}\frac{\partial^2}{\partial\xi_{\mathbf{k}}^2}-\frac{M_a}{2}\,\omega_{\mathbf{k}}^2\xi_{\mathbf{k}}^2+\frac{\hbar}{2}\,\omega_{\mathbf{k}} \quad (3.8a)$$

$$a_{\mathbf{k}}a_{\mathbf{k}}^+ = \frac{\hbar^2}{2M_a}\frac{\partial^2}{\partial\xi_{\mathbf{k}}^2}-\frac{M_a}{2}\,\omega_{\mathbf{k}}^2\xi_{\mathbf{k}}^2-\frac{\hbar}{2}\,\omega_{\mathbf{k}} \quad (3.8b)$$

and Eq. 3.6 becomes

$$\left(a_{\mathbf{k}}^+a_{\mathbf{k}}-\frac{\hbar}{2}\,\omega_{\mathbf{k}}\right)\underset{\mathbf{k}}{g_{r,s}}(\boldsymbol{\xi}_{\mathbf{k}}) = -\mathscr{E}_{r,\mathbf{k}}\underset{\mathbf{k}}{g_{r,s}}(\boldsymbol{\xi}_{\mathbf{k}}) \quad (3.9)$$

Multiplying both sides by $a_{\mathbf{k}}$, we get

$$a_{\mathbf{k}} a_{\mathbf{k}}^{+} a_{\mathbf{k}} \underset{\mathbf{k}}{g_{r,s}}(\boldsymbol{\xi}_{\mathbf{k}}) = -\left(\mathscr{E}_{r,\mathbf{k}} - \frac{\hbar}{2}\,\omega_{\mathbf{k}}\right) a_{\mathbf{k}} \underset{\mathbf{k}}{g_{r,s}}(\boldsymbol{\xi}_{\mathbf{k}})$$

By use of Eq. 3.8*b*, this can be rewritten in the form

$$\left(\frac{\hbar^2}{2M_a}\frac{\partial^2}{\partial \xi_{\mathbf{k}}^2} - \frac{M_a}{2}\,\omega_{\mathbf{k}}^2 \xi_{\mathbf{k}}^2\right) a_{\mathbf{k}} \underset{\mathbf{k}}{g_{r,s}}(\boldsymbol{\xi}_{\mathbf{k}}) = -(\mathscr{E}_{r,\mathbf{k}} - \hbar\omega_{\mathbf{k}}) a_{\mathbf{k}} \underset{\mathbf{k}}{g_{r,s}}(\boldsymbol{\xi}_{\mathbf{k}}) \qquad (3.10)$$

showing that $a_{\mathbf{k}} \underset{\mathbf{k}}{g_{r,s}}(\boldsymbol{\xi}_{\mathbf{k}})$ is also a solution of Eq. 3.6, with the corresponding eigenvalue $(\mathscr{E}_{r,\mathbf{k}} - \hbar\omega_{\mathbf{k}})$. Knowing one of the eigenfunctions $\underset{\mathbf{k}}{g_{r,s}}$ and the corresponding eigenvalue $\mathscr{E}_{r,\mathbf{k}}$, we can list all other eigenfunctions and eigenvalues as follows:

Eigenfunctions	*Eigenvalues*
$(a_{\mathbf{k}}^{+})^n \underset{\mathbf{k}}{g_{r,s}}(\boldsymbol{\xi}_{\mathbf{k}})$	$\mathscr{E}_{r,\mathbf{k}} + n\hbar\omega_{\mathbf{k}}$
$\underset{\mathbf{k}}{g_{r,s}}(\boldsymbol{\xi}_{\mathbf{k}})$	$\mathscr{E}_{r,\mathbf{k}}$
$(a_{\mathbf{k}})^n \underset{\mathbf{k}}{g_{r,s}}(\boldsymbol{\xi}_{\mathbf{k}})$	$\mathscr{E}_{r,\mathbf{k}} - n\hbar\omega_{\mathbf{k}}$

Each time the operator $a_{\mathbf{k}}$ operates on $\underset{\mathbf{k}}{g_{r,s}}(\boldsymbol{\xi}_{\mathbf{k}})$, the energy eigenvalue decreases by $\hbar\omega_{\mathbf{k}}$; hence, after a high enough number of successive operations, we would obtain negative values. These, however, must always be positive, since

$$\begin{aligned}\mathscr{E}_{r,\mathbf{k}} &= -\frac{\hbar^2}{2M_a}\int g^* \frac{\partial^2}{\partial \xi^2} g\, d\xi_{\mathbf{k}} + \frac{M_a}{2}\,\omega_{\mathbf{k}}^2 \int g^* \xi_{\mathbf{k}} g\, d\xi_{\mathbf{k}} \\ &= +\frac{\hbar^2}{2M_a}\int \frac{\partial g^*}{\partial \xi_{\mathbf{k}}}\frac{\partial g}{\partial \xi_{\mathbf{k}}}\, d\xi_{\mathbf{k}} + \frac{M_a}{2}\,\omega_{\mathbf{k}}^2 \int g^* \xi_{\mathbf{k}} g\, d\xi_{\mathbf{k}}\end{aligned}$$

In order to avoid a contradiction, we must postulate that one of the eigenfunctions $(a_{\mathbf{k}})^n \underset{\mathbf{k}}{g_{r,s}}(\boldsymbol{\xi}_{\mathbf{k}})$ becomes identically zero; if $g(\boldsymbol{\xi}_{\mathbf{k}})$ is the last function that is not identically zero, we have

$$a_{\mathbf{k}} g = \left(\frac{\hbar^2}{2M_a}\right)^{1/2} \frac{\partial g}{\partial \xi_{\mathbf{k}}} + \left(\frac{M_a \omega_{\mathbf{k}}}{2}\right)^{1/2} \xi_{\mathbf{k}} g = 0 \qquad (3.11)$$

Straightforward integration of Eq. 3.11 gives

$$g = C \exp\left(-\frac{M_a \omega_{\mathbf{k}}}{2\hbar}\,\xi_{\mathbf{k}}^2\right)$$

and the corresponding eigenvalue, obtained by replacing $g(\xi_{\mathbf{k}})$ by the above value in Eq. 3.6, is $\mathscr{E} = \frac{1}{2}\hbar\omega_{\mathbf{k}}$, so that the eigenvalues of Eq. 3.6 are of the well-known form

$$\mathscr{E}_{r,\mathbf{k}} = (n + \tfrac{1}{2})\hbar\omega_{\mathbf{k}}$$

We shall now show in a simple example that the "harmonic" theory developed above is such a poor approximation that it yields results in striking contradiction to the most common observations.

Let us consider a rod of some crystalline material, and assume that both ends are kept at different temperatures T_1 and T_2 ($T_1 < T_2$).

An upper limit for the heat flux carried by phonons of energy $\hbar\omega_{\mathbf{k}}$ traveling from the hot to the cold section is $r_{\mathbf{k}}c_{\mathbf{k}}\hbar\omega_{\mathbf{k}}$, where $r_{\mathbf{k}}$ is the number of phonons of energy $\hbar\omega_{\mathbf{k}}$, given by Bose-Einstein statistics: $r_{\mathbf{k}}\alpha[\exp(\hbar\omega_{\mathbf{k}}/kT) - 1]^{-1}$, and $c_{\mathbf{k}}$ the group velocity of these phonons, which is practically independent of temperature.

The net heat flux from the hot to the cold section is thus

$$\Delta Q = Q_{2\to 1} - Q_{1\to 2} = \sum_{\mathbf{k}} \hbar\omega_{\mathbf{k}}[r_{\mathbf{k}}(T_2) - r_{\mathbf{k}}(T_1)]$$

In other words, ΔQ appears to depend only on T_1 and T_2, regardless of the temperature gradient, a conclusion that is obviously wrong.

The anharmonic terms in the potential, and other perturbations that can act as phonon scatterers, have been neglected in our treatment, and their neglect is responsible for the shortcomings of the harmonic theory.

4. Introduction to Phonon Scattering

In the previous sections we have separated the variables in the Schrödinger equation for a crystal; following the procedure outlined by Born and Oppenheimer, we had to make a few approximations in our potential $V(\mathbf{R}_a)$ and to disregard some terms introducing an error of the order of m/M_a in the calculated energy.

The corrective terms in the Hamiltonian H can be introduced as a perturbation that will induce transitions between the unperturbed levels.

The standard methods of the theory of perturbations will presently be used to find an expansion of the eigenvalues of Eq. 3.1, including anharmonic terms.

The exact Hamiltonian H can be written as

$$H = H_0 + U \tag{4.1}$$

where H_0 includes only the harmonic terms in the potential, the terms of higher orders being taken up in the perturbation U, which is assumed to be small compared to H_0.

Since the eigenfunctions $g_s{}^0$ of H_0 form a complete orthonormal set, it is possible to expand the perturbed eigenfunctions g_s in terms of the unperturbed functions:

$$g_s = \sum_i b_{s,i} g_s{}^0 \tag{4.2}$$

in which the coefficients $b_{s,i}$ are all very small except $b_{s,s}$, which is close to unity.

Inserting the expansion of Eq. 4.2 in the Schrödinger equation

$$H g_s = E_s g_s \tag{4.3}$$

taking the inner product of each term by $g_j{}^*$, and using the orthonormal properties of the two sets of wave functions, we find

$$\sum_i b_{s,i}[(E_i{}^0 - E_s)\delta_{ij} + U_{ij}] = 0 \tag{4.4}$$

where δ_{ij} is the usual Kronecker symbol and $U_{ij} = \int g_j{}^{0*} U g_i{}^0 \, d^3\mathbf{R}$.

Equation 4.4 can be rewritten as

$$b_{s,j}[E_j{}^0 - E_s + U_{jj}] + \sum_{i \neq j} b_{s,i} U_{ij} = 0 \tag{4.5}$$

Letting $j = s$ and sorting terms by increasing order, we get from Eq. 4.5

$$E_s = E_s{}^0 + U_{ss} - \sum_{i \neq s} b_{s,i} U_{is} \tag{4.6}$$

If U_{ss} is different from zero, a first approximate eigenvalue of E_s is $(E_s{}^0 + U_{ss})$. However, in the case of an anharmonic perturbation of the third order corresponding to a cubic potential, U_{ss} vanishes for reasons of symmetry. Hence a perturbation calculation must be carried to the second order. The coefficients of the second-order terms in Eq. 4.6 are found in the following way. In Eq. 4.5 when $j \neq s$, we use the following approximations:

$$\begin{cases} \sum_{i \neq j} b_{s,i} U_{ij} & \sim U_{sj} \\ E_j{}^0 - E_s + U_{jj} & \sim E_j{}^0 - E_s{}^0 \end{cases}$$

This gives $b_{s,j} = -U_{sj}/(E_j{}^0 - E_s{}^0)$, so that the expansion of E_s to the second order, in the case of a cubic perturbation, is

$$E_s = E_s{}^0 + \sum_{i \neq s} \frac{U_{si}U_{is}}{E_i{}^0 - E_s{}^0} \tag{4.7}$$

If an energy level $E_r{}^0$ is initially degenerate, the perturbation removes the degeneracy by an amount

$$\Delta E_r{}^0 = \int g_{r,0}{}^* U g_{r,0} \, d^3\mathbf{R}_a$$

so that the cubic terms alone cannot split the degeneracy.

The transition probability produced by the perturbation between nondegenerate levels can be calculated by time-dependent perturbation theory. If $E_r{}^0$ and $E_s{}^0$ are the initial and final nonperturbed states, and U_{rs} the matrix element of the perturbation between these states, the probability $P_{r,s}(t)$ that a transition has occurred within a time t after applying the perturbation can be written

$$P_{r,s}(t) = 4\sin^2 \frac{(E_r{}^0 - E_s{}^0)t}{2\hbar} \frac{U_{rs}{}^2}{(E_r{}^0 - E_s{}^0)^2} \rho(E_s) \tag{4.8}$$

where $\rho(E_s)$ is the density of states with energy E_s. If we deal with a discrete set of states instead of a continuum, $\rho(E_s)$ must be replaced by $\delta(E - E_s)$, where δ is the Dirac delta function.

In order to find the transition probabilities arising from the third-order terms in $\mathscr{V}$ (see Eq. 3.2), we have to calculate the summation $\frac{1}{6} \sum_{a,b,c} B_{a,b,c} u_a u_b u_c$ in terms of the $\xi_{\mathbf{k}}$'s. This can be done by using Eq. 3.3, and the result is

$$\frac{1}{6} \sum_{a,b,c} B_{a,b,c} u_a u_b u_c = \frac{1}{6} \sum_{\mathbf{k},\mathbf{k}',\mathbf{k}''} b(\mathbf{k}, \mathbf{k}', \mathbf{k}'') \xi_{\mathbf{k}} \xi_{\mathbf{k}'} \xi_{\mathbf{k}''} \tag{4.9}$$

with

$$b(\mathbf{k}, \mathbf{k}', \mathbf{k}'') = \sum_{a,b,c} B_{a,b,c} \exp\left[i(\mathbf{k}\cdot\mathbf{R}_a + \mathbf{k}'\cdot\mathbf{R}_b + \mathbf{k}''\cdot\mathbf{R}_c)\right]$$

At this point, we should note that if all the atoms of the crystal, which is assumed to be infinite, are displaced by a lattice vector $\mathbf{R}$, the potential $\mathscr{V}(\mathbf{R}_a)$, which has the symmetry properties of the lattice, does not change. Since this operation multiplies each of the exponential terms in $b(\mathbf{k},\mathbf{k}',\mathbf{k}'')$ by $\exp\left[i(\mathbf{k} + \mathbf{k}' + \mathbf{k}'')\cdot\mathbf{R}\right]$, the condition for the cubic potential to remain invariant is

$$\mathbf{k} + \mathbf{k}' + \mathbf{k}'' = \mathbf{K} \tag{4.10}$$

where $\mathbf{K}$ is a lattice vector of the reciprocal lattice (see Sec. 5).

With use of Eqs. 3.7*a* and 3.7*b*, $\xi_{\mathbf{k}}$ can be written as

$$\xi_{\mathbf{k}} = (a_{\mathbf{k}} - a_{\mathbf{k}}^{+})\sqrt{2/(M_a\omega_{\mathbf{k}}^2)}$$

Introducing this value in the product $\xi_{\mathbf{k}}\xi_{\mathbf{k}'}\xi_{\mathbf{k}''}$ of Eq. 4.9 leads to a sum of the form

$$a_{\mathbf{k}}a_{\mathbf{k}'}a_{\mathbf{k}''} + a_{\mathbf{k}}a_{\mathbf{k}'}a_{\mathbf{k}''}^{+} + a_{\mathbf{k}}a_{\mathbf{k}'}^{+}a_{\mathbf{k}''} + a_{\mathbf{k}}^{+}a_{\mathbf{k}'}a_{\mathbf{k}''} + a_{\mathbf{k}}^{+}a_{\mathbf{k}'}^{+}a_{\mathbf{k}''} + a_{\mathbf{k}}^{+}a_{\mathbf{k}'}a_{\mathbf{k}''}^{+} + a_{\mathbf{k}}a_{\mathbf{k}'}^{+}a_{\mathbf{k}''}^{+} + a_{\mathbf{k}}^{+}a_{\mathbf{k}'}^{+}a_{\mathbf{k}''}^{+}$$

The matrix element between an initial (i) and a final (f) configuration is a combination of integrals of the type

$$\int (f)^* a_{\mathbf{k}}a_{\mathbf{k}'}a_{\mathbf{k}''}(i)\, d^3\xi_{\mathbf{k}} \tag{4.11}$$

and since the operators $a_{\mathbf{k}}$, $a_{\mathbf{k}'}$, and $a_{\mathbf{k}''}$ operate independently in the eigenfunctions $g_{\mathbf{k}}(\xi_{\mathbf{k}})$, $g_{\mathbf{k}'}(\xi_{\mathbf{k}'})$, and $g_{\mathbf{k}''}(\xi_{\mathbf{k}''})$, respectively, the integral (Eq. 4.11) reduces to the product of three independent integrals of the type

$$\int g_{p,s}{}_{\mathbf{k}}^{*} a_{\mathbf{k}} g_{r,s}{}_{\mathbf{k}}\, d\xi_{\mathbf{k}} = \sqrt{r_{\mathbf{k}}}\,\delta_{r_{\mathbf{k}}+1,p_{\mathbf{k}}} \tag{4.12}$$

or

$$\int g_{p,s}{}_{\mathbf{k}}^{*} a_{\mathbf{k}}^{+} g_{r,s}{}_{\mathbf{k}}\, d\xi_{\mathbf{k}} = \sqrt{r_{\mathbf{k}} + 1}\,\delta_{r_{\mathbf{k}}-1,p_{\mathbf{k}}} \tag{4.13}$$

As a result of the above discussion, the matrix element corresponding to the sum in Eq. 4.9 becomes

$$\frac{1}{6}\sum_{\mathbf{k}\mathbf{k}'\mathbf{k}''}{}' b(\mathbf{k}, \mathbf{k}', \mathbf{k}'') \times (\text{8 integrals of the type described in Eqs. 4.12 and 4.13}) \tag{4.14}$$

The prime in the summation indicates the condition of conservation of crystal momentum: $\mathbf{k} + \mathbf{k}' + \mathbf{k}'' = \mathbf{K}$.

The transition probability between discrete states, according to Eq. 4.8, is

$$P_{if} = 4\,\frac{\sin^2 (E_f - E_i)t}{(E_f - E_i)^2}\, U_{if}^2\, \delta(E_f - E_i)$$

It is zero unless

$$E_f - E_i = 0 \tag{4.15}$$

Since the initial state consists of a phonon of wave vector **k** about to collide with a phonon of wave vector **k**′, and assuming that the final state consists of a phonon of momentum **k**″, we have

$$\hbar(\omega_{\mathbf{k}} + \omega_{\mathbf{k}'} - \omega_{\mathbf{k}''}) = 0 \tag{4.16}$$

We can now understand the physical meaning of Eqs. 4.10 and 4.16. The first one deals with the conservation of crystal momentum; if

$\mathbf{K} = 0$, it is similar to the conservation of momentum in classical mechanics. Otherwise, we have a new situation referred to as an "Umklapp process" (Fig. 4.1), which actually permits the decay of

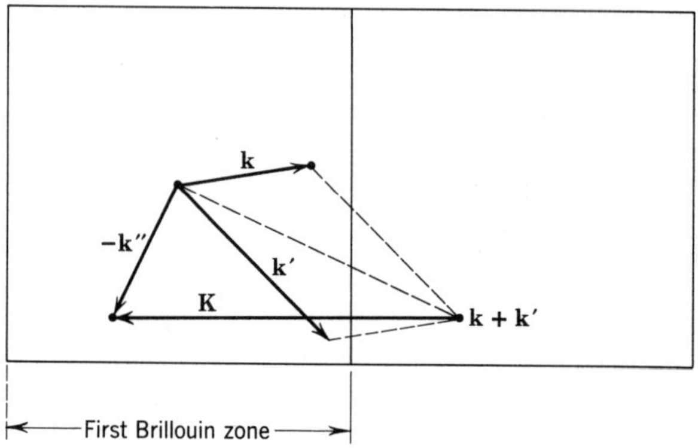

Fig. 4.1. "Umklapp" process.

momentum in the absence of a thermal gradient. The significance of this relation will become clearer after the study of Sec. 5. If the extremity of the vector sum $(\mathbf{k} + \mathbf{k}')$ falls outside the Brillouin zone, it is always possible to find a vector $(-\mathbf{k}'')$, deduced from $(\mathbf{k} + \mathbf{k}')$ by a translation along a lattice vector $\mathbf{K}$ of the reciprocal lattice, so that $-\mathbf{k}''$ (and hence $\mathbf{k}''$) is in the Brillouin zone. Equation 4.10 shows that the scattering of a phonon of wave vector $\mathbf{k}$ by a phonon of wave vector $\mathbf{k}'$ can create a phonon of wave vector $\mathbf{k}''$.

The latter relation (Eq. 4.16) corresponds to the classical result of perturbation theory, namely, that the only real transitions are those which conserve the energy. Other sources of phonon scattering, of great importance in the theory of heat conduction, are the local mass fluctuations resulting from the presence of isotopes or impurities in the lattice. If the masses of all the nuclei are not the same, the Hamiltonian can be written

$$H = \sum_a \left[-\frac{\hbar^2}{2} \overline{\left(\frac{1}{M_a}\right)} \nabla_a{}^2 + V(\mathbf{R}_a) \right] + \sum_a -\frac{\hbar^2}{2} \left[\frac{1}{M_a} - \overline{\left(\frac{1}{M_a}\right)} \right] \nabla_a{}^2 \quad (4.17)$$

where the second term can be considered as a perturbation and $\overline{1/M_a}$ is the average of the inverse mass.

In the Laplacian operator $\nabla_a{}^2$, the derivatives with respect to the displacements can be replaced by those with respect to $\xi_{\mathbf{k}}$ by using Eq. 3.3.

From Eqs. 3.7*a* and 3.7*b*, it results that

$$\frac{\partial}{\partial \xi_{\mathbf{k}}} = (a_{\mathbf{k}} + a_{\mathbf{k}}^{+}) \sqrt{2\bar{M}_a/\hbar}$$

or

$$\frac{\partial^2}{\partial \xi_{\mathbf{k}}^2} = 2(a_{\mathbf{k}}^2 + a_{\mathbf{k}}a_{\mathbf{k}}^{+} + a_{\mathbf{k}}^{+}a_{\mathbf{k}} + a_{\mathbf{k}}^{+2})\bar{M}_a/\hbar$$

and the matrix elements of the perturbation are of the form

$$\int \left[g_{r,s}^{*}{}_{\mathbf{k}} a_{\mathbf{k}}^2 g_{p,s}{}_{\mathbf{k}} + g^{*}_{r,s}{}_{\mathbf{k}} a_{\mathbf{k}}^{+2} g_{p,s}{}_{\mathbf{k}} + g^{*}_{r,s}{}_{\mathbf{k}} (a_{\mathbf{k}}^{+}a_{\mathbf{k}} + a_{\mathbf{k}}a_{\mathbf{k}}^{+}) g_{p,s}{}_{\mathbf{k}} \right] d\xi_{\mathbf{k}}$$

Using Eqs. 4.12 and 4.13, we see that the number $r_{\mathbf{k}}$ of phonons in the mode **k** changes by ± 2 or 0 in a scattering process caused by mass fluctuation.

The calculation of these matrix elements leads to an expression for the probability $P_{r,p}$ that a phonon in the state r undergoes a transition to the state p. Then the probability that a phonon undergoes a transition from the state r to *any* other state p is

$$P_r = \sum_p P_{r,p}$$

The mean free path Λ of the phonons, and the heat conductivity due to them,

$$K_{ph} = \tfrac{1}{3} C_v c \Lambda$$

can be estimated. Detailed calculations show that the scattering due to isotopes and/or small concentration of impurities is far from negligible. In other words, isotopes or impurities reduce drastically the heat conductivity by phonons. This conclusion is of great importance for thermoelectric applications in which one seeks scattering centers that are more efficient phonon scatterers than electron scatterers.

5. Reciprocal Lattice—Brillouin Zone

We have seen in Sec. 3 that the displacement $\mathbf{u}_a$ of the nuclei in a lattice can be expressed as a summation of sinusoidal plane waves of frequency $\nu_{\mathbf{k}} = \omega_{\mathbf{k}}/2\pi$ moving in the direction $\mathbf{k}$:

$$\mathbf{u}_a = \sum_{\mathbf{k}} \boldsymbol{\xi}_{\mathbf{k}} \exp\,(i\mathbf{k}\cdot\mathbf{R}_{a,0} + \omega_{\mathbf{k}}t) \tag{5.1}$$

The properties of these traveling waves are greatly clarified by introducing the concept of the reciprocal lattice. Given a lattice L of lattice vectors $\mathbf{L}_1$, $\mathbf{L}_2$, $\mathbf{L}_3$, the lattice l, reciprocal of L, is defined by its lattice vectors $\mathbf{l}_1$, $\mathbf{l}_2$, $\mathbf{l}_3$, such that

$$\mathbf{l}_i\cdot\mathbf{L}_j = 2\pi\,\delta_{ij} \tag{5.2}$$

As we shall see, this introduces a great simplification in the calculation of the summation 5.1. If we consider two vectors $\mathbf{k}$ and $\mathbf{k}'$ which differ from one another by a translation of the reciprocal lattice

$$\mathbf{k}' = \mathbf{k} + m\mathbf{l}_1 + n\mathbf{l}_2 + p\mathbf{l}_3 \tag{5.3}$$

(m, n, p being integers), $\exp\,(i\mathbf{k}'\cdot\mathbf{R}_{a0})$ can be written

$$\exp\,(i\mathbf{k}'\cdot\mathbf{R}_{a0}) = \exp\,(i\mathbf{k}\cdot\mathbf{R}_{a0})\exp\,[i\mathbf{R}_{a0}\cdot(m\mathbf{l}_1 + n\mathbf{l}_2 + p\mathbf{l}_3)] \tag{5.4}$$

where

$$\mathbf{R}_{a0} = M\mathbf{L}_1 + N\mathbf{L}_2 + P\mathbf{L}_3 \tag{5.5}$$

(M, N, P also being integers).

Replacing $\mathbf{R}_{a0}$ in Eq. 5.4 by its expression 5.5, and using Eq. 5.2, we get

$$\exp\,(i\mathbf{k}'\cdot\mathbf{R}_{a0}) = \exp\,(i\mathbf{k}\cdot\mathbf{R}_{a0})\exp\,[2\pi i(mM + nN + pP)] \tag{5.6}$$

but $mM + nN + pP$ is an integer, the second factor of Eq. 5.6 is therefore unity, and

$$\exp\,(i\mathbf{k}'\cdot\mathbf{R}_{a0}) = \exp\,(i\mathbf{k}\cdot\mathbf{R}_{a0})$$

We conclude that phonons whose wave vectors differ by a vector of the reciprocal lattice are equivalent. Hence, it is sufficient to know the characteristics of phonons with any wave vector $\mathbf{k}$ contained in a zone Z such that any other vector $\mathbf{k}'$ is in a zone Z' deduced from Z by a translation $m\mathbf{l}_1 + n\mathbf{l}_2 + p\mathbf{l}_3$.

The basic zone Z can be chosen in several ways, but the most convenient for our purpose is that of the "Brillouin zone," which is the volume included inside the mediator planes of the lines joining the origin to the nearest reciprocal lattice points (Fig. 5.1). This zone is often called the first Brillouin zone, as other zones (second, third, . . .) can be constructed in a similar way with the next nearest lattice points; we will use only the first zone, which shall be called the "Brillouin zone."

Consider a crystal with g atoms per unit cell. To each of the three degrees of freedom of the g atoms of the unit cell corresponds a normal mode of frequency $\nu_{\mathbf{k}} = \omega_{\mathbf{k}}/2\pi$; therefore, for each value of $\mathbf{k}$, we have $3g$ values of $\omega_{\mathbf{k}}$, which are the roots of a secular equation of degree $3g$. For small $\mathbf{k}$ (long wave lengths) it is a good approximation to take only

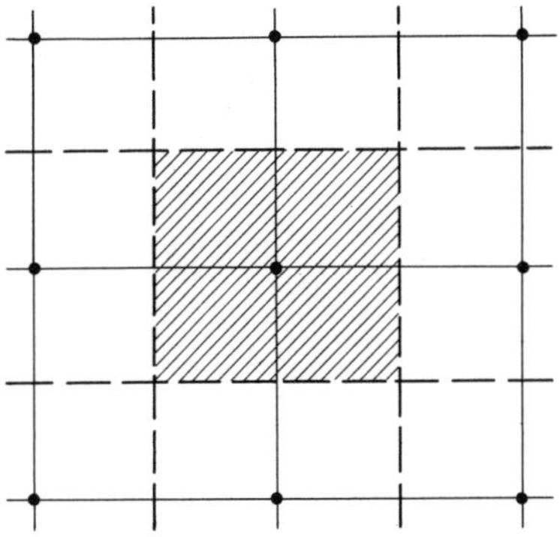

Fig. 5.1. The first Brillouin zone in a two-dimensional square lattice.

the first term of the series expansion of $\omega_{\mathbf{k}}(\mathbf{k})$ in terms of $k = |\mathbf{k}|$ along each branch.

a. Acoustical modes: Three of the roots can be expanded as

$$\omega_{\mathbf{k}} = Ak + \cdots$$

These represent the three low-frequency "acoustical" modes (one longitudinal and two transverse), in which the atoms vibrate in phase, and the disturbance corresponds to an elastic wave.

b. Optical modes: $3(g - 1)$ of the roots are of the form

$$\omega(k) = \omega(0) - Bk^2 + \cdots$$

In this case, the g atoms of the unit cell do not vibrate in phase and produce a varying electric dipole. These modes can be easily coupled with electromagnetic radiation, usually in the infrared range.

For illustration, we have represented on Fig. 5.2 the normal modes as a function of k_x for the case $g = 2$, giving three acoustical and three optical modes.

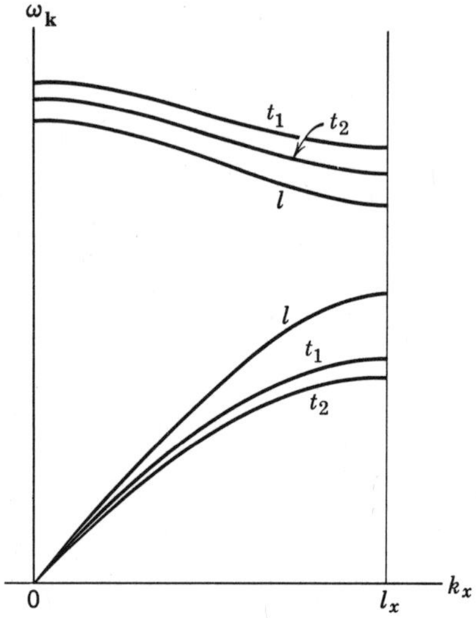

Fig. 5.2. The normal modes of a three-dimensional lattice having two atoms per unit cell.

6. Formulation of the Electronic Wave Equations

In Sec. 2, the eigenvalues $\mathscr{E}_s$ and eigenfunctions ψ_s of the electronic Hamiltonian H_e were supposed to be known.

Actually, the problem of solving Eq. 2.4 is extremely complex. The simplest thing to do is to consider the electrons as moving in a potential that is produced by the atomic cores slightly displaced from their average position and by all the valence electrons.

In order to solve Eq. 2.4, we must not disregard the spin co-ordinates of the electrons, and we can try to write ψ_s as a product of functions for the individual electrons, just as was done in Sec. 3 for the phonons:

$$\psi_s = \prod_i \phi_i(\mathbf{q}_i) = \prod_i \phi_i(\mathbf{r}_i, \sigma_i) \tag{6.1}$$

where $\mathbf{q}_i$ includes both the space and spin co-ordinates of electron i.

The Hamiltonian H_e can be written as

$$H_e = \sum_i H_i + \frac{1}{2} \sum_{i,j}{}' \frac{e^2}{r_{ij}} \tag{6.2}$$

where H_i includes the kinetic energy of electron i and its potential energy in the field of the nuclei, and the last term represents the Coulomb interaction between electrons i and j. The factor $\frac{1}{2}$ is needed to count each term only once, since $r_{ij} = r_{ji}$.

The exact wave function ψ_s is such that the quantity

$$E = \int \psi_s^* H_e \psi_s \, d\tau_1 \cdots d\tau_N$$

has its lowest possible value.

The Hartree Method

If the wave function ψ_s and the Hamiltonian H_e as given by Eqs. 6.1 and 6.2, respectively, are used, E becomes

$$E = \sum_i \phi_i^*(\mathbf{q}_i) H_i \phi_i(\mathbf{q}_i) \, d\tau_i + \tfrac{1}{2} \sum_{i,j}{}' \int \phi_i^*(\mathbf{q}_i)\phi_j^*(\mathbf{q}_j) \frac{e^2}{r_{ij}} \phi_i(\mathbf{q}_i)\phi_j(\mathbf{q}_j) \, d\tau_i \, d\tau_j \tag{6.3}$$

Equation 6.3 can be rewritten in the form

$$E = \sum_i \mathscr{E}_i$$

where

$$\mathscr{E}_i = \int \phi_i^*(\mathbf{q}_i) \left[H_i + \tfrac{1}{2} \sum_{j \neq i} \int |\phi_j(\mathbf{q}_j)|^2 \frac{e^2}{r_{ij}} \, d\tau_j \right] \phi_i(\mathbf{q}_i) \, d\tau_i$$

The sum in the brackets is the Hamiltonian for an electron moving in the potential of the nuclei and of the other electrons, and the integral is minimum when $\phi_i(\mathbf{q}_i)$ is the ground-state eigenfunction of this Hamiltonian.

Each of the optimum $\phi_i(\mathbf{q}_i)$ is thus obtained by solving the set of integrodifferential equations

$$\left[H_i + \tfrac{1}{2} \sum_{j \neq i} \int |\phi_j(\mathbf{q}_j)|^2 \frac{e^2}{r_{ij}} \, d\tau_j \right] \phi_i(\mathbf{q}_i) = \mathscr{E}_i \phi_i(\mathbf{q}_i) \tag{6.4}$$

Since it is impossible to solve this system directly, successive approximations should be used. We can choose a reasonable set of initial ϕ's, calculate from these the integrals $\int |\phi_j(\mathbf{q}_j)|^2 e^2/r_{ij}\, d\tau_j$ in terms of one co-ordinate only, find the eigenfunctions of the Hamiltonians by the use of perturbation theory, use these eigenfunctions to calculate the new Hamiltonians, and so on, until self-consistency is obtained between the eigenfunctions used in the Hamiltonians and those obtained by solving them.

This method has been used extensively in atomic problems and, to a certain extent, has been applied to solids. It should be noticed, however, that the function ψ_s, which is a product of one-electron functions, is not antisymmetric with respect to the exchange of two electrons, as it should be according to the Pauli exclusion principle. It can be shown that this drawback of the Hartree scheme is relatively unimportant in the case of isolated atoms but very severe in the case of solids.

The Hartree-Fock Method

A suitable antisymmetric wave function can be written as a Slater determinant:

$$\psi_s = (N!)^{-1/2} \begin{vmatrix} \phi_1(\mathbf{q}_1) & \phi_1(\mathbf{q}_2) \cdots \phi_1(\mathbf{q}_N) \\ \phi_2(\mathbf{q}_1) & \phi_2(\mathbf{q}_2) \cdots \phi_2(\mathbf{q}_N) \\ \cdot & \cdot \qquad\quad \cdot \\ \cdot & \cdot \qquad\quad \cdot \\ \cdot & \cdot \qquad\quad \cdot \\ \phi_N(\mathbf{q}_1) & \phi_N(\mathbf{q}_2) \cdots \phi_N(\mathbf{q}_N) \end{vmatrix} \tag{6.5}$$

It is readily seen that the interchange of two variables in Eq. 6.5 merely changes the sign of ψ_s. If this new function is used in the integral $E = \int \psi_s^* H_e \psi_s \, d\tau_1 \cdots d\tau_N$, we obtain an expansion similar to Eq. 6.3, except that new "exchange" terms of the form

$$\mathscr{E}_{\text{exc}} = \int \phi_k^*(\mathbf{q}_i)\phi_{k'}^*(\mathbf{q}_j) \frac{e^2}{r_{ij}} \phi_k(\mathbf{q}_j)\phi_{k'}(\mathbf{q}_i) \, d\tau_1 \cdots d\tau_N$$

are now introduced.

By the same reasoning as in the Hartree scheme, we obtain for the ϕ's a new set of equations

$$\left\{H_1 + \sum_j \int |\phi_j(\mathbf{q}_2)|^2 \frac{e^2}{r_{12}} \, d\tau_2 - \sum_{\substack{j \\ \text{spin } j \,=\, \text{spin } i}} \int \phi_j^*(\mathbf{q}_2)\phi_i(\mathbf{q}_2) \frac{e^2}{r_{12}} \, d\tau_2\right\} \phi_i(\mathbf{q}_1) = \mathscr{E}\phi_i(\mathbf{q}_1) \tag{6.6}$$

In principle, this system can be solved as previously; however, the actual calculations become inextricable before self-consistency can be obtained, and drastic approximations will have to be made. We shall see in the next section that the symmetry of a crystal permits a simplification of the problem.

The readers interested in a more thorough discussion of the Hartree and Hartree-Fock schemes are referred to Slater's remarkable review article on the electronic structure of solids, which is mentioned in the general bibliography.

7. Solution of the Electronic Wave Equation—Bloch Functions

Effect of Crystal Symmetry

In this section, we shall investigate how the symmetry of the crystal affects the choice of the wave functions and the nature of the approximation scheme. The translation operator, an essential mathematical tool for this study, will first be introduced.

It is easy to verify that, if two operators commute, their eigenfunctions are proportional. Keeping this in mind, let us define an operator $A_{\mathbf{L}}$ in such a way that

$$A_{\mathbf{L}}\phi(\mathbf{r}) = \phi(\mathbf{r} + \mathbf{L}) \tag{7.1}$$

where $\phi(\mathbf{r})$ is an eigenfunction of H. If $A_{\mathbf{L}}$ now operates on the product $v(\mathbf{r})\phi(\mathbf{r})$, where $v(\mathbf{r})$ has the periodicity of the lattice, we see that

$$A_{\mathbf{L}}v(\mathbf{r})\phi(\mathbf{r}) = v(\mathbf{r})A_{\mathbf{L}}\phi(\mathbf{r}) \tag{7.2}$$

showing that $A_{\mathbf{L}}$ and $v(\mathbf{r})$ commute. But $H = -\hbar^2/2m\nabla^2 + v(\mathbf{r})$, and, as $A_{\mathbf{L}}$ commutes also with ∇^2, it commutes with H. Consequently,

$$A_{\mathbf{L}}\phi(\mathbf{r}) = \lambda\phi(\mathbf{r}) \tag{7.3}$$

λ being a constant factor of proportionality.

We can write this factor in the form $\exp(\boldsymbol{\alpha}\cdot\mathbf{L})$ and $\phi(\mathbf{r})$ in the form $\exp(\boldsymbol{\alpha}\cdot\mathbf{r})\, s_{\alpha}(\mathbf{r})$, where $s_{\alpha}(\mathbf{r})$ is a function of $\mathbf{r}$ which we shall now specify.

With these notations, Eq. 7.3 becomes

$$\begin{aligned} A_{\mathbf{L}} \exp(\boldsymbol{\alpha}\cdot\mathbf{r})s_{\alpha}(\mathbf{r}) &= \exp(\boldsymbol{\alpha}\cdot\mathbf{L}) \exp(\boldsymbol{\alpha}\cdot\mathbf{r})s_{\alpha}(\mathbf{r}) \\ &= \exp[\boldsymbol{\alpha}\cdot(\mathbf{r}+\mathbf{L})]s_{\alpha}(\mathbf{r}) \end{aligned}$$

and, by virtue of Eq. 7.1,

$$s_{\alpha}(\mathbf{r}) = s_{\alpha}(\mathbf{r}+\mathbf{L})$$

showing that $s_{\alpha}(\mathbf{r})$ is a periodic function of $\mathbf{r}$, having the periodicity of the lattice.

Furthermore, the amplitude of the electron wave must remain finite anywhere at infinity. This requires that $\boldsymbol{\alpha}$ be a pure imaginary vector. Finally, an appropriate wave function can be written as

$$\phi_{\mathbf{k}}(\mathbf{r}) = \exp(i\mathbf{k}\cdot\mathbf{r})s_{\mathbf{k}}(\mathbf{r}) \tag{7.4}$$

This important form of the electron wave is known as the Bloch wave.

Expansion in Terms of Bloch Functions

Suppose that we know the wave functions for $\mathbf{k} = 0$, and that $s_0(\mathbf{r})$ is the one which corresponds to the ground state of the electron under consideration.

In order to see how $s_{\mathbf{k}}(\mathbf{r})$ depends on $\mathbf{k}$, we can try to approximate $\phi_{\mathbf{k}}(\mathbf{r})$ by $\exp(i\mathbf{k}\cdot\mathbf{r})s_0(\mathbf{r})$.

The energy associated with this trial function is

$$W = \int \exp(-i\mathbf{k}\cdot\mathbf{r})s_0^*(\mathbf{r})H \exp(i\mathbf{k}\cdot\mathbf{r})s_0(\mathbf{r})\, d\tau$$

where

$$H = -\frac{\hbar^2}{2m}\nabla^2 + \mathscr{V}(\mathbf{r})$$

But it is easy to see that

$$\begin{aligned} \nabla^2 \exp(i\mathbf{k}\cdot\mathbf{r})s_0(\mathbf{r}) = \exp(i\mathbf{k}\cdot\mathbf{r})\nabla^2 s_0(\mathbf{r}) &+ 2i\exp(i\mathbf{k}\cdot\mathbf{r})\mathbf{k}\cdot\boldsymbol{\nabla}s_0(\mathbf{r}) \\ &+ k^2\exp(i\mathbf{k}\cdot\mathbf{r})s_0(\mathbf{r}) \end{aligned}$$

so that

$$\begin{aligned} W = \int s_0^*(\mathbf{r})Hs_0(\mathbf{r})\, d\tau - i\frac{\hbar^2}{m}\mathbf{k}\cdot \int s_0^*(\mathbf{r})\boldsymbol{\nabla}s_0(\mathbf{r})\, d\tau \\ + \frac{\hbar^2}{2m}k^2 \int s_0^*(\mathbf{r})s_0(\mathbf{r})\, d\tau \end{aligned}$$

or

$$W = E_0 + \frac{\hbar^2 k^2}{2m}$$

E_0 being the eigenvalue associated with $s_0(\mathbf{r})$. This shows that the energy associated with the approximate wave function exp $(i\mathbf{k}\cdot\mathbf{r})s_0(\mathbf{r})$ depends on $\mathbf{k}$ in the same way as the plane wave exp $(i\mathbf{k}\cdot\mathbf{r})$.

In other words, as long as the extremity of $\mathbf{k}$ is not too close to the edge of the Brillouin zone, the isoenergy surface on which this extremity falls is nearly spherical, and exp $(i\mathbf{k}\cdot\mathbf{r})s_0(\mathbf{r})$ is a fairly good approximation to the actual wave function.

In order to see how good the approximation is, let us consider the actual wave equation

$$H \exp (i\mathbf{k}\cdot\mathbf{r})s_{\mathbf{k}}(\mathbf{r}) = \mathscr{E}_k \exp (i\mathbf{k}\cdot\mathbf{r})s_{\mathbf{k}}(\mathbf{r}) \tag{7.5}$$

and multiply both sides by exp $(-i\mathbf{k}\cdot\mathbf{r})$. We get

$$\exp (-i\mathbf{k}\cdot\mathbf{r})H \exp (i\mathbf{k}\cdot\mathbf{r})s_{\mathbf{k}}(\mathbf{r}) = \mathscr{E}_k s_{\mathbf{k}}(\mathbf{r})$$

By expanding the term ∇^2 exp $(i\mathbf{k}\cdot\mathbf{r})s_{\mathbf{k}}(\mathbf{r})$ as before, we obtain

$$\left[-\frac{\hbar^2}{2m}(\nabla_i{}^2 + 2i\mathbf{k}\cdot\boldsymbol{\nabla}_i - k^2) + \mathscr{V}(\mathbf{r})\right]s_{\mathbf{k}}(\mathbf{r}) = \mathscr{E}_k s_{\mathbf{k}}(\mathbf{r})$$

or

$$\left(H - \frac{\hbar^2}{m}i\mathbf{k}\cdot\boldsymbol{\nabla}\right)s_{\mathbf{k}}(\mathbf{r}) = \left(\mathscr{E}_k - \frac{\hbar^2 k^2}{2m}\right)s_{\mathbf{k}}(\mathbf{r}) \tag{7.6}$$

For $\mathbf{k} = 0$, we get, as expected,

$$Hs_0(\mathbf{r}) = \mathscr{E}_0 s_0(\mathbf{r})$$

Regarding the term $-\hbar^2/m\, i\mathbf{k}\cdot\boldsymbol{\nabla}$ as a perturbation, we can use the results of second-order perturbation theory to find $\mathscr{E}_{0,k}$ in terms of $k = |\mathbf{k}|$ and of the eigenvalues for $k = 0$.

Actually,

$$-\frac{\hbar^2}{m}i\mathbf{k}\cdot\boldsymbol{\nabla} = \frac{\hbar}{m}\mathbf{k}\cdot\mathbf{p}$$

where $\mathbf{p}$ is the momentum operator. The energy of the perturbation is

$$\Delta\mathscr{E} = \left(\mathscr{E}_{0,k} - \frac{\hbar^2 k^2}{2m}\right) - \mathscr{E}_0 = \sum_n{}' \frac{H'_{0n}H'_{n0}}{\mathscr{E}_0 - \mathscr{E}_n}$$

where n refers to the nth band, and

$$H'_{0n} = \frac{\hbar}{m}\mathbf{k}\cdot \int s_{0,0}(\mathbf{r})\mathbf{p}s_{0,n}(\mathbf{r})\, d\tau$$

The integral can be abbreviated as $\langle\mathbf{p}\rangle_{0n}$ so that

$$H'_{0n} = \frac{\hbar}{m}\mathbf{k}\cdot\langle\mathbf{p}\rangle_{0n} = \frac{\hbar}{m}\alpha_n k\langle p\rangle_{0n}$$

where

$$\alpha_n = \cos(\mathbf{k}, \langle\mathbf{p}\rangle_{0n})$$

With these notations, we thus have

$$\mathscr{E}_{0,k} = \mathscr{E}_0 + \frac{\hbar^2k^2}{2m} + \frac{\hbar^2k^2}{m^2}\sum_n{}' \alpha_n \frac{\langle p\rangle_{0n}\langle p\rangle_{n0}}{\mathscr{E}_0 - \mathscr{E}_n} \tag{7.7}$$

or

$$\mathscr{E}_{0,k} = \mathscr{E}_0 + \frac{\hbar^2k^2}{m}\left[1 + \frac{2}{m}\sum_n{}' \alpha_n \frac{\langle p\rangle_{0n}\langle p\rangle_{n0}}{\mathscr{E}_0 - \mathscr{E}_n}\right]$$

The quantity between brackets is often larger than 1.

By defining a new mass m^* in such a way that

$$\frac{1}{m^*} = \frac{1}{m}\left[1 + \frac{2}{m}\sum_n{}' \alpha_n \frac{\langle p\rangle_{0n}\langle p\rangle_{n0}}{\mathscr{E}_0 - \mathscr{E}_n}\right] \tag{7.8}$$

the expression for $\mathscr{E}_{0,k}$ becomes similar to that for a plane wave.

Equation 7.8 actually defines the "effective mass" m^*, at the center of the Brillouin zone. The calculation of m^* has been carried out in a few cases where the function $s_0(\mathbf{r})$ was known. For the alkali metals, for instance, m^*/m is slightly smaller than 1, except for lithium, for which $m^*/m \sim 1.4$.

In Sec. 8, we shall see that an effective-mass tensor can be defined at any point of the Brillouin zone which is a minimum of the conduction band. The merit of the above calculation is that it shows clearly that the effective mass is closely related to the shape of the energy band around $\mathbf{k} = 0$, and gives a direct evaluation of m^*.

Remark. The above perturbation treatment has defined $\mathscr{E}_k$ as a continuous function of $|\mathbf{k}|$.

Actually, in a finite crystal, containing a finite number of electrons, the number of states is finite, so that $\mathscr{E}_k$ is not, strictly speaking, a continuous function of $\mathbf{k}$. The smallest transition corresponds to

$$\Delta E_{\min} = E_1 - E_0 \simeq \frac{\hbar^2}{2m}[(k + \Delta k_{\min})^2 - k^2]$$

where $\Delta k_{\min} = 2\pi/Na$ for a crystal of lattice constant a containing N atoms. For a typical crystal of 1 cm^3, $\Delta E_{\min}$ is of the order of 10^{-12} ev.

Boundary Conditions

The propagation of phonon and electron waves in a bounded crystal depends not only on the properties of the infinite lattice but also on the boundary conditions, which may considerably increase the complexity of the problem by mixing all types of wave functions.

To avoid the mathematical complications that would arise with the introduction of arbitrary boundary conditions, Born and von Kármán conceived the idea of cyclic boundary conditions. The ideal crystal introduced by them has a large number of atomic planes in each direction beyond which it repeats itself exactly as if the original crystal had been translated by a distance $\mathscr{L}$ equal to one of its dimensions (Fig. 7.1).

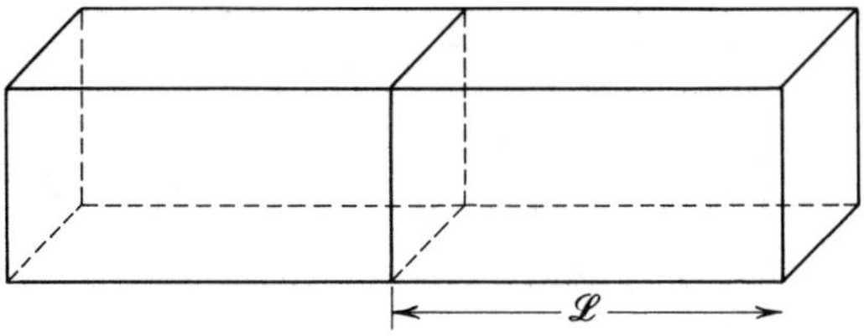

Fig. 7.1. The cyclic boundary conditions.

If there are M atomic planes in the distance $\mathscr{L}$, and $\mathscr{L}$ is parallel to $\mathbf{L}_1$, we have

$$\mathscr{L} = M\mathbf{L}_1$$

Our original crystal will now have a periodicity with period $M\mathbf{L}_1$ in the $\mathbf{L}_1$ direction, $N\mathbf{L}_2$ in the $\mathbf{L}_2$ direction, and $P\mathbf{L}_3$ in the $\mathbf{L}_3$ direction. The introduction of the arbitrary periodicity ensures that the wave functions are identical at opposite faces of the crystal.

For instance, the Bloch wave function $\exp(i\mathbf{k}\cdot\mathbf{r})s_{\mathbf{k}}(\mathbf{r})$ at one point of a face must be invariant on changing $\mathbf{r}$ into $\mathbf{r} + M\mathbf{L}_1$. By virtue of Eq. 6.10, this means

$$\exp(i\mathbf{k}\cdot\mathbf{r}) \equiv \exp(i\mathbf{k}\cdot\mathbf{r})\ \exp(i\mathbf{k}\cdot M\mathbf{L}_1)$$

hence

$$\exp(i\mathbf{k}\cdot M\mathbf{L}_1) \equiv 1$$

or

$$\mathbf{k}\cdot\mathbf{L}_1 = |k_1||L_1| = 2\pi\frac{n}{M}$$

where k_1 is the projection of $\mathbf{k}$ along $\mathbf{L}_1$, and n is an arbitrary integer.

We know from the definition of the reciprocal lattice that

$$\mathbf{l}_1\cdot\mathbf{L}_1 = |l_1||L_1| = 2\pi$$

It follows that

$$|k_1| = \frac{n}{M}|l_1|$$

with similar relations for the two other components of **k**. This means that the extremity of the **k** vectors are the lattice nodes of a three-dimensional array with lattice constants

$$\frac{\mathbf{l}_1}{M}, \quad \frac{\mathbf{l}_2}{N}, \quad \frac{\mathbf{l}_3}{P}$$

The volume of the unit cell of this new reciprocal lattice, that is, the reciprocal density of **k** vectors in **k**-space, is

$$\sum = \frac{(\mathbf{l}_1 \times \mathbf{l}_2)\cdot\mathbf{l}_3}{MNP} = \frac{(2\pi)^3}{MNP(\mathbf{L}_1 \times \mathbf{L}_2)\cdot\mathbf{L}_3} = \frac{8\pi^3}{V}$$

V being the volume of the cyclic crystal.

8. Introduction to Conduction Theory—Wannier Formalism and the Effective Mass

We shall now suppose that an electric field which can be considered as a perturbation is applied to the crystal. We shall investigate the effect of this perturbation on an insulator containing one extra electron and a fixed positive charge to maintain neutrality. This will enable us to discuss the electrical conduction in semiconductors, under the reasonable assumption that mutual interactions between electrons remain negligible. In the case of metals, these mutual interactions will have to be taken into consideration.

The perturbed configuration takes a relatively simple form if the unperturbed eigenfunctions, which form a complete set when **k** is limited to the Brillouin zone, are expanded in Fourier series in **k**-space in the following way:

$$\psi_{\mathbf{k}}(\mathbf{r}) = N^{-1/2} \sum_{\mathbf{R}_p} W_n(\mathbf{r} - \mathbf{R}_p) \exp(i\mathbf{k}\cdot\mathbf{R}_p) \tag{8.1}$$

The number N, entering the normalization constant, is the number of atoms in the crystal.

The functions $W_n(\mathbf{r} - \mathbf{R}_p)$, known as the Wannier functions, are localized in space, and behave somewhat as atomic wave functions.

Equation 8.1 is essentially similar to the expression for a wave packet localizing an electron; in the latter formulation, the momentum p of the electron loses its meaning, while in the former one, the wave vector $\mathbf{k}$ is no more explicit.

Each Wannier function is, however, characteristic of a given band, identified by the subscript n.

The Wannier functions can be calculated by multiplying both sides of Eq. 8.1 by $\exp(-i\mathbf{k}\cdot\mathbf{R}_q)$ and summing over all $\mathbf{k}$'s in the Brillouin zone:

$$\sum_{\mathbf{k}} \exp(-i\mathbf{k}\cdot\mathbf{R}_q)\psi_{\mathbf{k}}(\mathbf{r}) = N^{-1/2} \sum_{\mathbf{R}_p} W_n(\mathbf{r} - \mathbf{R}_p) \sum_{\mathbf{k}} \exp[i\mathbf{k}\cdot(\mathbf{R}_p - \mathbf{R}_q)]$$

The sum over $\mathbf{k}$ in the right-hand side is

$$\begin{array}{ll} 0 & \text{for} \quad \mathbf{R}_p \neq \mathbf{R}_q \\ N & \text{for} \quad \mathbf{R}_p = \mathbf{R}_q \end{array}$$

Hence,

$$\sum_{\mathbf{k}} \exp(-i\mathbf{k}\cdot\mathbf{R}_q)\psi_{\mathbf{k}}(\mathbf{r}) = N^{1/2} W_n(\mathbf{r} - \mathbf{R}_q) \tag{8.2}$$

Let us prove now that the functions W form an orthonormal set. From Eq. 8.2, we have

$$\int W_n^*(\mathbf{r} - \mathbf{R}_p) W_{n'}(\mathbf{r} - \mathbf{R}_q)\, d^3\mathbf{r} = N^{-1} \sum_{\mathbf{k},\mathbf{k}'} \exp[i(\mathbf{k}\cdot\mathbf{R}_p - \mathbf{k}'\cdot\mathbf{R}_q)] \int \psi_{\mathbf{k}}^*(\mathbf{r})\psi_{\mathbf{k}'}(\mathbf{r})\, d^3\mathbf{r}$$

but $\psi_{\mathbf{k}}(\mathbf{r})$ and $\psi_{\mathbf{k}'}(\mathbf{r})$, eigenfunctions of the unperturbed Hamiltonian, are orthonormal, so that

$$\int \psi_{\mathbf{k}}^*(\mathbf{r})\psi_{\mathbf{k}'}(\mathbf{r})\, d^3\mathbf{r} = \delta_{\mathbf{k},\mathbf{k}'}$$

and

$$\int W_n^*(\mathbf{r} - \mathbf{R}_p) W_{n'}(\mathbf{r} - \mathbf{R}_q)\, d^3\mathbf{r} = N^{-1} \sum_{\mathbf{k}} \exp[i\mathbf{k}\cdot(\mathbf{R}_p - \mathbf{R}_q)]$$

For the same reason as above, this is $\delta_{\mathbf{R}_p,\mathbf{R}_q}$, and consequently, the Wannier functions form an orthonormal set derived from $\psi_{\mathbf{k}}(\mathbf{r})$ by a unitary transformation whose coefficients are $\exp(i\mathbf{k}\cdot\mathbf{R}_p)$.

Let $E_{\mathbf{k}}$ be the eigenvalue corresponding to $\psi_{\mathbf{k}}$:

$$H_0\psi_{\mathbf{k}} = E_{\mathbf{k}}\psi_{\mathbf{k}} \tag{8.3}$$

If in Eq. 8.3 we replace $\psi_{\mathbf{k}}$ by its expansion, given in Eq. 8.1, multiply both sides by $\exp(-i\mathbf{k}\cdot\mathbf{R}_q)$, and sum over all $\mathbf{k}$, we get

$$\sum_{\mathbf{k}} \exp(-i\mathbf{k}\cdot\mathbf{R}_q) H_0 \sum_{\mathbf{R}_p} N^{-1/2} \exp(i\mathbf{k}\cdot\mathbf{R}_p) W_n(\mathbf{r}-\mathbf{R}_p) = \sum_{\mathbf{k}} \exp(-i\mathbf{k}\cdot\mathbf{R}_q) E_{\mathbf{k}} \sum_{\mathbf{R}_p} N^{-1/2} \exp(i\mathbf{k}\cdot\mathbf{R}_p) W_n(\mathbf{r}-\mathbf{R}_p)$$

By changing the order of the summations and noting that H_0 operates only on the Wannier functions, we can reduce the above relation to

$$H_0 W_n(\mathbf{r}-\mathbf{R}_q) = \sum_{\mathbf{R}_p} \mathscr{E}(\mathbf{R}_p-\mathbf{R}_q) W_n(\mathbf{r}-\mathbf{R}_p) \tag{8.4}$$

where

$$\mathscr{E}(\mathbf{R}_i) \equiv N^{-1} \sum_{\mathbf{k}} E_{\mathbf{k}} \exp(i\mathbf{k}\cdot\mathbf{R}_i)$$

and hence,

$$\int W_n^*(\mathbf{r}-\mathbf{R}_p) H_0 W_n(\mathbf{r}-\mathbf{R}_q)\, d^3\mathbf{r} = \mathscr{E}(\mathbf{R}_p-\mathbf{R}_q) \tag{8.5}$$

Thus, $\mathscr{E}(\mathbf{R}_p-\mathbf{R}_q)$ is the nondiagonal matrix element of H_0 between $W_n(\mathbf{r}-\mathbf{R}_p)$ and $W_n(\mathbf{r}-\mathbf{R}_q)$. The matrix element of H_0 between two Wannier functions corresponding to different bands (n and $n' \neq n$) is zero.

Perturbed Problem

We are now ready to consider the problem of a lattice perturbed, for instance, by an electric field.

The Hamiltonian of the perturbed problem is

$$H = H_0 + H_1 \tag{8.6}$$

where H_1 is the perturbation.

The eigenfunctions of the perturbed Hamiltonian can be expanded in terms of the eigenfunctions of the unperturbed Hamiltonian, which in turn can be expanded in terms of the Wannier functions introduced above. Thus, we can expand the perturbed eigenfunctions in terms of the Wannier functions as a double sum

$$\psi = \sum_{n,\mathbf{R}_j} U_n(\mathbf{R}_j) W_n(\mathbf{r}-\mathbf{R}_j) \tag{8.7}$$

over the Wannier functions corresponding to all the bands (n) and all the atoms of the crystal ($\mathbf{R}_j$).

Using the standard techniques of perturbation theory to calculate the coefficients $U_n(\mathbf{R}_j)$, we can write

$$(H_0+H_1)\psi = \sum_{n,\mathbf{R}_j} (H_0+H_1) U_n(\mathbf{R}_j) W_n(\mathbf{r}-\mathbf{R}_j) = E \sum_{n,\mathbf{R}_j} U_n(\mathbf{R}_j) W_n(\mathbf{r}-\mathbf{R}_j) \tag{8.8}$$

where E is the energy of the perturbed system. Multiplying both sides of Eq. 8.8 by $W_m(\mathbf{r} - \mathbf{R}_i)$, integrating over all space, and using Eq. 8.5, we have

$$\sum_{m,\mathbf{R}_j} [\mathscr{E}_n(\mathbf{R}_i - \mathbf{R}_j)\,\delta_{nm} + V_{nm}(\mathbf{R}_i, \mathbf{R}_j)]U_m(\mathbf{R}_j) = EU_n(\mathbf{R}_j) \tag{8.9}$$

where

$$V_{nm}(\mathbf{R}_i, \mathbf{R}_j) = \int W_m(\mathbf{r} - \mathbf{R}_i)H_1W_n(\mathbf{r} - \mathbf{R}_j)\,d^3\mathbf{r} \tag{8.10}$$

$V_{nm}(\mathbf{R}_i, \mathbf{R}_j)$ is the matrix element of H_1 between two Wannier functions corresponding to the mth and nth bands, and centered on atoms i and j.

The set of equations labeled Eq. 8.9 gives an exact solution of our problem. The problem is greatly simplified if the nondiagonal matrix elements of $H_1(n \neq m)$ can be neglected, and this simplification is legitimate in a few important cases, namely, when the energy difference ΔE between the top of the valence band and the bottom of the conduction band is appreciable.

Furthermore, we can also neglect the matrix elements $V_{nn}(\mathbf{R}_i, \mathbf{R}_j)$ of H_1 centered on two different lattice sites $(i \neq j)$, since the Wannier functions decrease very fast, away from the lattice site on which they are centered.

Finally, the set of equations labeled Eq. 8.9 reduces, in first approximation, to the set

$$\sum_{\mathbf{R}_j} [\mathscr{E}_n(\mathbf{R}_i - \mathbf{R}_j)U_n(\mathbf{R}_j)] + V_{nn}(\mathbf{R}_i)U(\mathbf{R}_i) = EU(\mathbf{R}_i) \tag{8.11}$$

The basic feature of the present treatment is that the coefficients $U_n(\mathbf{R})$ can actually be regarded as continuous functions $U_n(\mathbf{r})$, which take the values $U_n(\mathbf{R}_i)$ at the ith lattice point. For instance, in the absence of a perturbation (Eq. 8.1), $U_n(\mathbf{R}_i) = \exp(i\mathbf{k}\cdot\mathbf{R}_i)$ is the value of the continuous function $\exp(i\mathbf{k}\cdot\mathbf{r})$ of a plane wave at the ith lattice point; and such a plane wave is a solution of a Schrödinger equation for a free particle.

This suggests that the $U_n(\mathbf{R}_i)$ could be found as discrete values of the eigenfunctions of a Schrödinger-like equation for a particle in the presence of only the perturbation H_1.

In order to set up this equation, we shall try to convert the difference equation, Eq. 8.11, into a differential equation. For this purpose, we shall use the method suggested by Wannier and developed by Slater.

The propagation vector $\mathbf{k}$ can be related to a pseudomomentum $\mathbf{p}$ by the relation $\hbar\mathbf{k} = \mathbf{p}$.

Following the procedure used to set up the Schrödinger equation of a single particle, we replace $\mathbf{k}$ by the operator $-i\boldsymbol{\nabla}$ whenever it appears in $E(\mathbf{k})$.

If we replace k_x by $-i(\partial/\partial x)$ in the expansion

$$\exp(-ik_xR_x) = 1 - ik_xR_x - \tfrac{1}{2}k_x{}^2R_x{}^2 + \cdots$$

and let the result operate on a function $f(x)$, we get

$$\exp(-ik_xR_x)\,f(x) = f(x) - R_x\frac{\partial f}{\partial x} + \tfrac{1}{2}R_x{}^2\frac{\partial^2 f}{\partial x^2} + \cdots = f(x - R_x)$$

By a simple generalization, we can thus write

$$\exp(-i\mathbf{k}\cdot\mathbf{R}_j)U(\mathbf{r}) = U(\mathbf{r} - \mathbf{R}_j) \tag{8.12}$$

Therefore, using Eqs. 8.4 and 8.12, we find

$$\sum_{\mathbf{R}_j} \mathscr{E}_n(\mathbf{R}_j)U(\mathbf{r} - \mathbf{R}_j) = E(\mathbf{k})U(\mathbf{r}) \tag{8.13}$$

The left member of Eq. 8.13 is the same as that of Eq. 8.11, except that U is now a continuous function of $\mathbf{r}$, and $(\mathbf{R}_i - \mathbf{R}_j)$ is replaced by $\mathbf{R}_j$. Consequently, the function $U(\mathbf{r})$ is defined by

$$E(\mathbf{k})U(\mathbf{r}) + V_{n,n}(\mathbf{r})U(\mathbf{r}) = EU(\mathbf{r}) \tag{8.14}$$

Under the assumption that it is not degenerate at an extremum, $E(\mathbf{k})$ can be expanded in Taylor series around this extremum. By a proper choice of the axis, the mixed differentials appearing in the expansion vanish, and we get

$$E(\mathbf{k}) = E(\mathbf{k}_0) + \frac{1}{2}\left[\left(\frac{\partial^2 E}{\partial k_x{}^2}\right)_{k_0}(k_x - k_{0x})^2 + \left(\frac{\partial^2 E}{\partial k_y{}^2}\right)_{k_0}(k_y - k_{0y})^2 + \cdots\right] + (\text{third-order terms}) + \cdots \tag{8.15}$$

If this expansion, limited to the second-order terms, is introduced in Eq. 8.14, we obtain a second-order equation for $U(\mathbf{r})$. However, the neglect of the higher-order terms in Eq. 8.15 is justified only if the eigenfunction varies slowly with $\mathbf{r}$. This is not the case for $U(\mathbf{r})$, but it is the case for the function $\phi(\mathbf{r})$ defined by

$$\phi(\mathbf{r}) = \exp(-i\mathbf{k}_0\cdot\mathbf{r})U(\mathbf{r}) \tag{8.16}$$

Expressing Eq. 8.16 in terms of x, y, and z, we see that

$$\begin{aligned}[k_x - k_{x0}]\exp[i(k_{0x}x + k_{0y}y + k_{0z}z)]\phi(x,y,z) \\ &= -i\frac{\partial}{\partial x}\{\exp[i(k_{0x}x + \cdots)]\phi(x,y,z)\} \\ &\qquad - k_{0x}\exp[i(k_{0x}x + \cdots)]\phi(x,y,z) \\ &= -i\exp[i(k_{0x}x + \cdots)]\frac{\partial}{\partial x}\phi(x,y,z)\end{aligned}$$

and

$$[k_x - k_{x0}]^2 \exp[i(k_{0x}x + \cdots)]\phi(x, y, z)$$
$$= -\exp[i(k_{0x}x + \cdots)]\frac{\partial^2}{\partial x^2}\phi(x, y, z)$$

so that Eq. 8.16 transforms Eq. 8.14 into

$$E_{k0}\phi - \frac{1}{2}\left[\left(\frac{\partial^2 E}{\partial k_x^2}\right)_{k_0}\frac{\partial^2 \phi}{\partial x^2} + \cdots + \cdots\right] + V_{n,n}\phi = E_k\phi \tag{8.17}$$

Now we can introduce the notations

$$\left.\begin{aligned} \frac{1}{m_x} &= \hbar^2\left(\frac{\partial^2 E}{\partial k_x^2}\right)_{k_0} \\ \frac{1}{m_y} &= \hbar^2\left(\frac{\partial^2 E}{\partial k_y^2}\right)_{k_0} \\ \frac{1}{m_z} &= \hbar^2\left(\frac{\partial^2 E}{\partial k_z^2}\right)_{k_0} \end{aligned}\right\} \tag{8.17a}$$

When we use these notations, Eq. 8.17 becomes

$$-\frac{\hbar^2}{2}\left(\frac{1}{m_x}\frac{\partial^2 \phi}{\partial x^2} + \cdots + \cdots\right) + V_n(x, y, z)\phi = (E_k - E_{k_0})\phi \tag{8.18}$$

This equation is similar to the Schrödinger equation, except that the Laplacian operator ∇^2 operating on ϕ is now replaced by

$$\left(\frac{m}{m_x}\frac{\partial^2}{\partial x^2} + \frac{m}{m_y}\frac{\partial^2}{\partial y^2} + \frac{m}{m_z}\frac{\partial^2}{\partial z^2}\right)$$

The masses m_x, m_y, and m_z are the "effective masses" in the directions x, y, and z, respectively.

In Sec. 7, we introduced the isotropic effective mass m^* for $k_0 = 0$. As we see here, the effective mass is not, in general, a scalar but a tensor, of which m_x, m_y, and m_z are the diagonal components.

These quantities may be negative, for instance, near the top of the valence band. In order to grasp the significance of the whole concept of effective mass, we must remember that, according to the correspondence principle, the motion of a quantum-mechanical particle tends to coincide with the motion predicted by classical dynamics when states of high quantum numbers are considered. Hence, the "average" motion of an electron in a lattice under the influence of an electric field **E** can be described by

$$m^*\left(\frac{d\mathbf{v}}{dt} + \frac{\mathbf{v}}{\tau}\right) = -e\mathbf{E} \tag{8.19}$$

that is, by the classical equation of motion in which m^* has been substituted for the mass of the free electron. If m^* is negative, Eq. 8.19 describes the average motion of a particle of charge $+e$ and mass $|m^*|$.

The physical interpretation of this result is the following: Suppose we have an almost completely filled band in which an electron is missing; under an electric field, all the electrons will tend to make a transition to the next highest unoccupied state. This is equivalent to saying that the vacancy in the unoccupied state causes transition in the direction opposite to that of the electrons.

The above phenomena might be illustrated by a mechanical analog: Suppose a series of balls are laid on a vertical rim, as shown in Fig. 8.1.

Fig. 8.1. Mechanical analog of a hole.

We can say either that the balls tend to fall by gravity to occupy the vacancy; or that the vacancy tends to move up, and hence we must ascribe a negative mass to the vacancy.

The effective-mass approximation developed above holds if:

a. The change of energy introduced by the perturbation is small.
b. The perturbing potential is a slowly varying function of position.
c. The band structure of the crystal is nondegenerate at the point where we are making our series expansion.

We shall now discuss the special cases of the electrons and holes in silicon and germanium.

Electrons in Silicon

The energy band has a minimum for $|\mathbf{k}| = k_0 \sim \frac{2}{3}K_{100}$ along the (100) and equivalent crystallographic directions.

Consequently, the points of minimum energy are

	I	II	III	IV	V	VI
k_x	$+k_0$	$-k_0$	0	0	0	0
k_y	0	0	$+k_0$	$-k_0$	0	0
k_z	0	0	0	0	$+k_0$	$-k_0$

around each minimum, the constant energy surface can be approximated by an ellipsoid of revolution, and the effective masses derived from the curvature of these ellipsoids are

$$m_l^* = (0.97 \pm 0.02)m$$
$$m_t^* = (0.10 \pm 0.01)m$$

Figure 8.2 represents a section of the energy surfaces made by the plane $k_z = 0$.

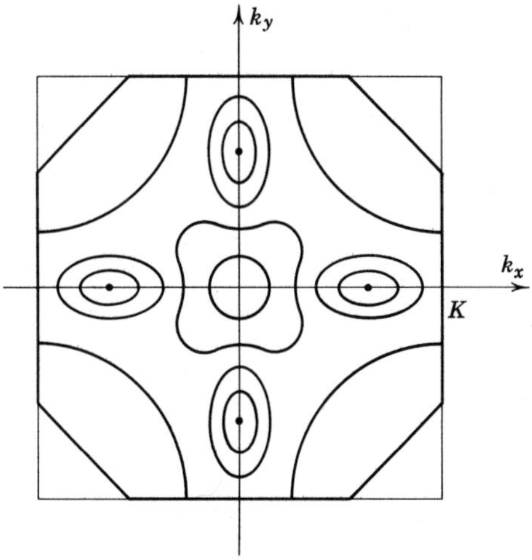

Fig. 8.2. Energy contours for silicon. Section by the plane $k_z = 0$.

Electrons in Germanium

The energy minima are on the edges of the Brillouin zone, along the (111) and equivalent directions.

Consequently, there are eight equivalent points of lowest energy, of co-ordinates $(\pm K, \pm K, \pm K)$.

Around each minimum, the surface of constant energy can be represented as a half ellipsoid of revolution, and the effective masses corresponding to the curvatures of the ellipsoids are

$$m_l^* = 1.58m$$
$$m_t^* = 0.082m$$

Figure 8.3 shows a section of the energy surfaces by the plane $k_y + k_z = 0$.

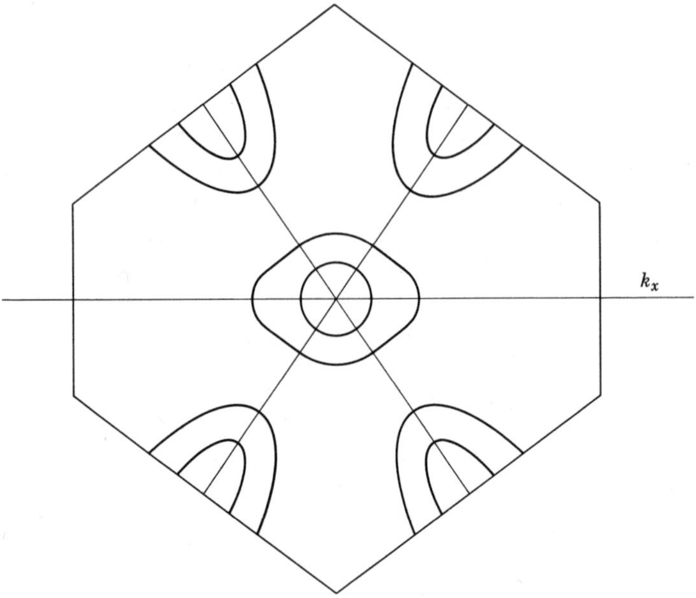

Fig. 8.3. Energy contours for germanium. Section by the plane $k_y + k_z = 0$.

Holes in Germanium and Silicon

In both germanium and silicon the maximum of the valence band actually happens to be at the center of the Brillouin zone, which from symmetry considerations is triply degenerate. The introduction of spin-orbit interaction partly removes this degeneracy, leaving the band doubly degenerate, however. The equation for $E(\mathbf{k})$ near the top of the valence band is given by the quadratic equation

$$E = \frac{-\hbar^2}{2m}[ak^2 \pm \sqrt{b^2k^4 - c^2(k_x^4 + k_y^4 + k_z^4)}] \tag{8.20}$$

and for the lower nondegenerate band by

$$E = -A - \frac{\hbar^2}{2m} ak^2 \tag{8.21}$$

where A represents the effect of spin-orbit interaction.

Cyclotron-resonance experiments give the following values to the parameters of Eqs. 8.20 and 8.21:

	Germanium	Silicon
a	13.0	4.1
b	12.01	2.83
c	7.1	2.33

Even if the upper limit of the valence band is degenerate so that the effective-mass approximation is not rigorously valid, it is common practice to associate an isotropic effective mass to each of the two degenerate bands. This is legitimate, since cyclotron-resonance experiments do not show appreciable anisotropy in the resonance peak associated with holes when the orientation of the sample in the cavity is changed. The values usually accepted for the average mass ratios of the holes are

	Germanium	Silicon
m_1/m	0.04	0.165
m_2/m	0.3	0.51

If an electric field is applied to a perfect crystal, the electrons in the conduction band as well as the holes in the valence band gain energy from the field until their wave vector eventually reaches the edge of the Brillouin zone. This happens, for instance, at $k = \pm 2\pi/a$ if the field is applied along one of the crystallographic directions. There, the Bragg conditions for total reflection are fulfilled, and k changes from $\pm 2\pi/a$ to $\mp 2\pi/a$. ($A \rightarrow B$ or $B \rightarrow A$ in Fig. 8.4.)

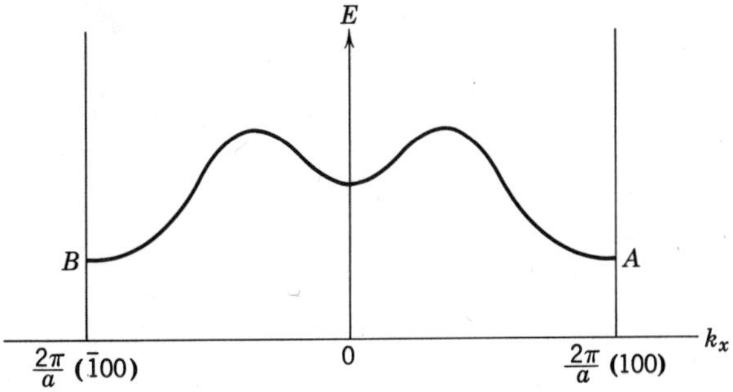

Fig. 8.4. Schematic band diagram in **k**-space, showing the possibility of Bragg reflection. (Transition $A \rightarrow B$.)

If the field is high enough, that is, of the order of 10^6 v/cm, so that the matrix element of H_1 between Wannier functions corresponding to different bands becomes appreciable, interband transitions are possible, as predicted by Zener.

Actually, lattice imperfections and other neglected perturbations that cause transitions between the stationary states usually prevent the electrons from gaining enough energy from the field to reach the top of the conduction band. The same is also true for the holes.

9. Statistical Mechanics of Electrons in Solids

Since electrons are particles of spin $\frac{1}{2}$, it results from the Pauli exclusion principle that they obey Fermi-Dirac statistics. Accordingly, the probability that an electron in equilibrium with a system at the temperature T occupies a state of energy E is

$$P(E) = \left[1 + g \exp\left(\frac{E - F}{kT}\right)\right]^{-1} \tag{9.1}$$

where g is the multiplicity of the state E, and F a parameter derived from the condition

$$\sum_E P(E) = N \tag{9.2}$$

which expresses that the sum of the probable populations of all energy states is equal to the total number of electrons. F has the dimension of an energy, and is known as the Fermi energy.

Curves showing $P(E)$ for various temperatures are indicated in Fig. 9.1.

If the energy levels form a continuum, with a density of states $\rho(E)$, Eq. 9.2 becomes

$$N = \int_0^\infty \rho(E)P(E)\,dE = \int_0^\infty \rho(E)\left[1 + g \exp\left(\frac{E - F}{kT}\right)\right]^{-1} dE \tag{9.3}$$

The origin in the energy scale has been chosen as the lowest energy level that can be occupied by the electrons under consideration.

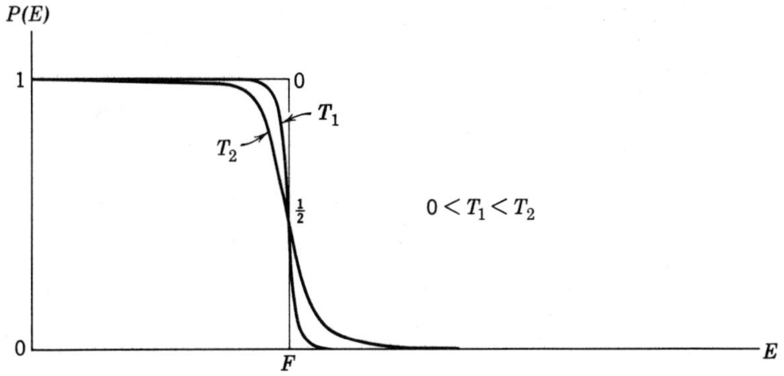

Fig. 9.1. The Fermi-Dirac distribution function.

When $E >> F$, $P(E) \simeq A \exp(-E/kT)$, so that Fermi statistics reduce to Maxwell-Boltzmann statistics in the limiting case of high E. For instance, in an electron gas of very low density (N small), electrons can occupy only a small fraction of the lowest available states, so that the exclusion principle is satisfied even if the electrons have a Maxwellian distribution. Experiment shows that this is the case for insulators and semiconductors having less than 10^{-5} conduction electrons per atom.

If the density of conduction electrons is much higher, $P(E)$ must be described by Eq. 9.1. This is, in particular, the case for metals, since they all have about one free electron per atom.

The statistical distributions that have been introduced are correct under the assumptions that:

a. The electron gas is in equilibrium with itself.

b. Any applied field is small compared to the crystalline field, and such that the energy gained from the field by an electron is small compared to the thermal energy.

c. The energy-band diagram is independent of the position in the crystal. This implies that the crystal must be uniform, in particular, free of *p-n* junctions or dislocations.

10. Elementary Notions of Transport Theory

The Boltzmann Equation

If external fields (electric and magnetic) are applied, the population of the electronic energy levels results from an equilibrium between upward transitions (energy gained from the field) and downward transitions (energy lost by collisions with phonons and impurities).

The Boltzmann equation

$$\left(\frac{\partial f}{\partial t}\right)_{\text{fields}} + \left(\frac{\partial f}{\partial t}\right)_{\text{collisions}} = 0 \tag{10.1}$$

merely expresses the steady state of the population $f(\mathbf{v})$ of states where the electrons have the velocity $\mathbf{v}$. Under the assumption that the collision rate is independent from the applied fields, both terms of Eq. 10.1 can be expressed simply in terms of the applied fields and of transition probabilities:

$$\left(\frac{\partial f}{\partial t}\right)_{\substack{\text{fields}\\ \mathbf{E}\text{ and }\mathbf{H}}} = \left[e\mathscr{E} + \frac{e}{c}(\mathbf{v} \times \mathbf{H})\right] \cdot \nabla f(\mathbf{v}) \tag{10.2a}$$

$$\left(\frac{\partial f}{\partial t}\right)_{\text{collisions}} = \sum_{\mathbf{v}'} T_{\mathbf{v}'\to\mathbf{v}} - \sum T_{\mathbf{v}\to\mathbf{v}'} \tag{10.2b}$$

The probability $T_{\mathbf{v}\to\mathbf{v}'}$ that an electron of velocity $\mathbf{v}$ acquires after collision the velocity $\mathbf{v}'$ is proportional to

$$f(\mathbf{v})[1 - f(\mathbf{v}')]M_{\mathbf{v}\mathbf{v}'}{}^2\rho(\mathbf{v})\rho(\mathbf{v}')\,dt$$

where $M_{\mathbf{v}\mathbf{v}'}$ is the matrix element of the scattering Hamiltonian between the state of velocity $\mathbf{v}$ and that of velocity $\mathbf{v}'$.

Since $f(\mathbf{v})$ and $f(\mathbf{v}')$ are small compared to unity, the complementary factors $(1 - f)$ are close to unity, and the transition probabilities are of the first order in $f(\mathbf{v})$. The terms arising from electron-electron collisions are quadratic, since they involve products of the form

$$f(\mathbf{v})f(\mathbf{v}')[1 - f(\mathbf{v})][1 - f(\mathbf{v}')]$$

The Boltzmann equation can give only a first-order approximation of the motion of electrons in solids under weak applied fields. The matrix element $M_{\mathbf{v}\mathbf{v}'}$ of the scattering Hamiltonian between the initial state of velocity $\mathbf{v}$ and the final state of velocity $\mathbf{v}'$ is calculated with the zero-order wave functions, which have the symmetry of the lattice. Application of a field can drastically change the symmetry of these wave functions. For instance, it has been shown by Landau that, if a magnetic field is applied in the z-direction, the electronic wave functions take the form $\exp(ik_z z) f(x, y)$ where $f(x, y)$ is a complicated function of x and y. The average motion of the electrons (from the correspondence principle) is circular in a plane perpendicular to the field. If the magnetic field is so high that the diameter of the trajectories is only of the order of the mean free path, the matrix elements $M_{\mathbf{v}\mathbf{v}'}$ obviously lose their physical significance. For slightly different reasons, the Boltzmann equation cannot adequately describe the conduction under high electric field.

Alternative Approach to Conduction Theory: The Energy-Gain Method

Let us approach the problem from another standpoint. Suppose that a system S is in thermal equilibrium with a heat reservoir Q, under the assumption that the interaction between S and Q is so weak that the respective energy levels are practically unperturbed. If S and Q both contain a sufficient number of particles, their energy levels form a continuum; and if the energy of the system decreases by a small amount ΔE, the same amount of energy is taken up by the heat reservoir.

Statistical mechanics states that, in a classical system having a continuous distribution of energy levels, the density of occupied states with energy between E and $(E + dE)$ is of the form

$$\rho(E)dE = A \exp\left(-\frac{E}{kT}\right) dE \tag{10.3}$$

where A is a normalizing constant.

The probability that the system loses the energy ΔE to the reservoir per unit time is

$$P(E, E - \Delta E) = \frac{2\pi}{\hbar} \rho(E) R(E) |M|^2 \tag{10.4}$$

where $R(E)$ is the density of states of the reservoir, and $2\pi/\hbar\, \rho(E)|M|^2$ the probability that one of the particles of the sample undergoes a transition from the energy E to $(E - \Delta E)$. The probability for the opposite transition is the same, in view of the time reversibility of Schrödinger's equation.

When thermal equilibrium is reached, the energy lost by the system per unit time must exactly cancel the energy gained by the reservoir, so that

$$P(E, E - \Delta E) = P(E - \Delta E, E) \tag{10.5}$$

or, by virtue of Eq. 10.4,

$$\rho(E)R(E)|M|^2 = \rho(E - \Delta E)R(E + \Delta E)|M|^2$$

If we now use the general results of Eq. 10.3, we have

$$\frac{R(E + \Delta E)}{R(E)} = \frac{\rho(E)}{\rho(E - \Delta E)} = \exp\left(-\frac{\Delta E}{kT}\right) \tag{10.6}$$

Thus, the favored transitions are those that decrease the energy of the system, and consequently, the population of the lower energy states must be the larger.

If now energy is fed by a battery into a system composed of electrons, these electrons gain, between the instant when the field is applied (supposedly at $t = -\infty$) and that of the measurement, an average energy δE from the field $\mathscr{E}$ in the sample.

We can now define a mean free path $\mathbf{l}$ such that

$$\delta E = -e\mathscr{E}\cdot\mathbf{l} \tag{10.7}$$

In this equation, $\mathbf{l}$ and δE are, in general, functions of the velocity of the electron. It is advantageous to use the concept of mean free time, rather than that of mean free path, because in most cases the mean free time does not vary much with the velocity of the particles. If we let $\|\tau\|$ be the mean-free-time tensor, we have

$$\mathbf{l} = \|\tau\|\mathbf{v} \tag{10.8}$$

In order to illustrate some of the basic concepts of conduction theory in a simple case, it will be instructive to discuss the case of a linear lattice before taking up the three-dimensional problem.

Case of a One-Dimensional Lattice

The crystalline lattice will play the role of the heat reservoir, and the electron gas will be the system undergoing transitions. For the sake of simplicity, we shall consider only elastic collisions (those that change the sign of the electron velocity), and we shall assume that the applied field is so weak that it does not appreciably change the value of this velocity.

Let P be the probability per unit time that such a collision occurs. The probability $p(n, t)$ that an electron undergoes n transitions ($\mathbf{v} \to -\mathbf{v}$) during a time t is, according to Poisson statistics,

$$p(n, t) = \frac{(Pt)^n \exp(-Pt)}{n!} \tag{10.9}$$

Let us now call $\mathscr{P}_+$ and $\mathscr{P}_-$ the probabilities that at a given instant the velocity of the electrons is positive or negative, respectively; the energy δE can be written

$$\delta E = e\mathscr{E} \int_{-\infty}^{0} (\mathscr{P}_+ v - \mathscr{P}_- v)\, dt \tag{10.10}$$

Since the laws of dynamics in an electromagnetic field are invariant with respect to time reversal (if all magnetic fields are reversed), Eq. 10.10 can be rewritten

$$\delta E = e\mathscr{E}v \int_{0}^{+\infty} (\mathscr{P}_+ - \mathscr{P}_-)\, dt \tag{10.11}$$

It is easy to see that

$$\mathscr{P}_+ = \sum_{n \text{ even}} p(n, t) \tag{10.12a}$$

and

$$\mathscr{P}_- = \sum_{n \text{ odd}} p(n, t) \tag{10.12b}$$

By use of Eqs. 10.12*a*, 10.12*b*, and 10.9, Eq. 10.11 takes the form

$$\delta E = e\mathscr{E}v \int_0^{\infty} \left[1 - \frac{Pt}{1} + \frac{(Pt)^2}{2!} - \frac{(Pt)^3}{3!} + \cdots\right] \exp(-Pt)\, dt \tag{10.13}$$

or

$$\delta E = e\mathscr{E}v \int_0^{\infty} \exp(-2Pt)\, dt = \frac{e\mathscr{E}v}{2P} \tag{10.14}$$

The mean free path, as defined by Eq. 10.7 is thus $l = v/2P$, and the mean free time τ is l/v, or $1/2P$.

It is possible, in principle, to solve the three-dimensional problem. However, the actual calculations usually cannot be carried through without making simplifying assumptions. Such simplified calculation will be outlined in the next section.

11. Calculation of the Current Density

We shall presently outline the calculation for the conductivity under the simplifying assumption that the mean free time and the effective mass are scalar quantities, and that the field is directed along the x-direction. We have in this case

$$\mathbf{l} = \tau(|\mathbf{v}|)\mathbf{v} \tag{11.1}$$

and

$$\delta E = -e\tau(|\mathbf{v}|)\mathbf{v}\cdot\boldsymbol{\mathscr{E}} = -e\tau\mathscr{E}v_x \tag{11.2}$$

The x-component of the current density is

$$j_x = Ne\int_0^\infty f(v)v_x\,d^3\mathbf{v} \tag{11.3}$$

where $f(v)$ is the distribution function for the electrons of energy $\frac{1}{2}m^*v^2 - \delta E$, which can be written

$$f(v) = A\exp\left(-\frac{m^*v^2}{2kT} + \frac{\delta E}{kT}\right) \tag{11.4}$$

A being the normalization constant.

The current density now takes the explicit form

$$j_x = Ne\frac{\displaystyle\int_0^\infty v_x\exp\left(\frac{m^*v^2}{2kT} + \frac{\delta E}{kT}\right)d^3\mathbf{v}}{\displaystyle\int_0^\infty \exp\left(-\frac{m^*v^2}{2kT} + \frac{\delta E}{kT}\right)d^3\mathbf{v}} \tag{11.5}$$

If the applied field is small, so that $\delta E/kT << 1$, $\exp(\delta E/kT)$ can be expanded to the first order, and j_x becomes

$$j_x = Ne\frac{\displaystyle\int v_x\left(1 + \frac{\delta E}{kT}\right)\exp\left(-\frac{m^*v^2}{2kT}\right)d^3\mathbf{v}}{\displaystyle\int\left(1 + \frac{\delta E}{kT}\right)\exp\left(-\frac{m^*v^2}{2kT}\right)d^3\mathbf{v}} \tag{11.6}$$

If the crystal is isotropic,

$$\langle v_x^2\rangle = \tfrac{1}{3}\langle v^2\rangle = \langle v^2/3\rangle$$

hence, we can assume that the quantities whose averages are equal are themselves close to each other:

$$v_x^2 \cong \tfrac{1}{3}v^2 \tag{11.7}$$

If we use Eq. 11.2, Eq. 11.7, and the usual expression for the volume element

$$d^3\mathbf{v} = 4\pi v^2\,dv$$

the current density becomes

$$j_x = \frac{Ne^2\mathscr{E}}{3kT}\,\frac{\displaystyle\int_0^\infty \tau v^4 \exp\left(-\frac{m^*v^2}{2kT}\right)dv}{\displaystyle\int_0^\infty v^2 \exp\left(-\frac{m^*v^2}{2kT}\right)dv} \tag{11.8}$$

If the denominator is transformed by use of the relation

$$\int_0^\infty m^*v^2 \exp\left(-\frac{m^*v^2}{2kT}\right)v^2\,dv = 3kT\int_0^\infty \exp\left(-\frac{m^*v^2}{2kT}\right)v^2\,dv$$

the current density takes the final form

$$j_x = \frac{Ne^2\mathscr{E}}{m^*}\,\frac{\displaystyle\int_0^\infty \tau v^4 \exp\left(-\frac{m^*v^2}{2kT}\right)dv}{\displaystyle\int_0^\infty v^4 \exp\left(-\frac{m^*v^2}{2kT}\right)dv} \tag{11.9}$$

The ratio of the two integrals in Eq. 11.9 is merely the average of τ weighted by the function $v^4 \exp(-m^*v^2/2kT)$, so that we can write

$$j_x = \frac{Ne^2\mathscr{E}}{m^*}\langle\tau\rangle \tag{11.10}$$

Accordingly, the conductivity is

$$\sigma = \frac{j_x}{\mathscr{E}} = \frac{Ne^2}{m^*}\langle\tau\rangle \tag{11.11}$$

and the mobility is

$$\mu = \frac{\sigma}{Ne} = \frac{e}{m^*}\langle\tau\rangle \tag{11.12}$$

For a metal, the principle of the method outlined above is still valid, but Fermi-Dirac statistics must be used.

It follows from the Fermi-Dirac distribution that only the electrons with energy close, within a few kT, to the Fermi energy contribute to the electrical and thermal conductivity. This explains why the electronic specific heat in solids is so small compared to the specific heat of the lattice, except near the absolute zero, where the thermal energy is not sufficient to excite lattice vibrations.

12. Scattering of Electrons by Phonons

We have seen in previous sections that the probability of transition $P(\mathbf{v}, \mathbf{v}')$ for an electron scattered by a phonon is proportional to the product $|M|^2\rho(E_\mathbf{v})R(E_\mathbf{k})$, where M is the matrix element for the transition, $\rho(E_\mathbf{v})$ the density of electronic states with velocity $\mathbf{v}$, and $R(E_\mathbf{k})$ the density of phonons of wave vector $\mathbf{k}$.

We shall now proceed to calculate these three factors separately.

Calculation of the Matrix Element

Let us refer to the initial state (before scattering) and the final state (after scattering) by the subscripts i and f, respectively.

The matrix element for the scattering can be written

$$M = \int \phi_f{}^* g_f{}^* U \phi_i g_i \, d\xi_a \cdots d^3\mathbf{r} \tag{12.1}$$

The ϕ's are the one-electron Bloch functions introduced in Sec. 7:

$$\phi = \exp{(i\mathbf{k}\cdot\mathbf{r})} s_\mathbf{k}(\mathbf{r}, \xi_a)$$

The g's are the nuclear wave functions introduced in Secs. 2 and 3, and U is the scattering potential.

We assume initially that the scattering by phonons is the main limitation for the mean free path of the carriers. We represent this scattering by a perturbing potential U.

The periodic part of the Bloch function can be expanded as a Fourier series in terms of $\mathbf{r}$:

$$s_\mathbf{k}(\mathbf{r}\cdots, \xi_a\cdots) = \sum_n B_n \exp{(i\mathbf{K}_n\cdot\mathbf{r})} \tag{12.2}$$

where $\mathbf{K}_n$ is a vector of the reciprocal lattice. Using this expansion in the matrix element M, we get

$$M = \sum_{n,m} B_n^* B_m \int g_f^*(\xi_a \cdots)|U|g_i(\xi_a \cdots) \times \exp\left[i(-\mathbf{k}_f - \mathbf{K}_n + \mathbf{k}_i + \mathbf{K}_m)\cdot\mathbf{r}\right] d\xi_a \cdots d^3\mathbf{r} \qquad (12.3)$$

If we suppose that U is not a periodic function of $\mathbf{r}$, each of the integrals in the summation (Eq. 12.3) is zero, except those for which

$$-\mathbf{k}_f - \mathbf{K}_n + \mathbf{k}_i + \mathbf{K}_m = 0 \qquad (12.4)$$

If we limit ourselves to the first Brillouin zone ($\mathbf{K}_n' = \mathbf{K}_m$), it follows that $M = 0$ unless $\mathbf{k}_f = \mathbf{k}_i$. In other words, the scattering is not effective. Consequently, to be effective, the scattering potential must be a periodic function of $\mathbf{r}$.

This periodic function can be written

$$U = W(\mathbf{r}, \xi) \exp(i\mathbf{k}_{ph}\cdot\mathbf{r}) \qquad (12.5)$$

where W is a periodic function of $\mathbf{r}$ and $\mathbf{k}_{ph}$ is the propagation vector of a phonon wave. If we use Eq. 12.5 for U in Eq. 12.3, we get

$$M \propto \int g_f^*(\xi \cdots) W(\mathbf{r}, \xi) \exp\left[i(\mathbf{k}_i - \mathbf{k}_f + \mathbf{k}_{ph})\cdot\mathbf{r}\right] \times g_i(\xi \cdots)\, d\xi_a \cdots d^3\mathbf{r} \qquad (12.6)$$

and the condition for M to be different from zero is

$$\mathbf{k}_f = \mathbf{k}_i + \mathbf{k}_{ph} \qquad (12.7)$$

which expresses the conservation of momentum.

Furthermore, the conservation of energy requires that $|\mathbf{k}_f| \approx |\mathbf{k}_i|$, since the reduced mass of the nuclei is much bigger than the electron mass. Consequently, the scattering process can be represented by Fig. 12.1, from which it is clear that

$$|\mathbf{k}_{ph}| = |\mathbf{k}_f - \mathbf{k}_i| = 2|\mathbf{k}_i| \sin \theta/2 \qquad (12.8)$$

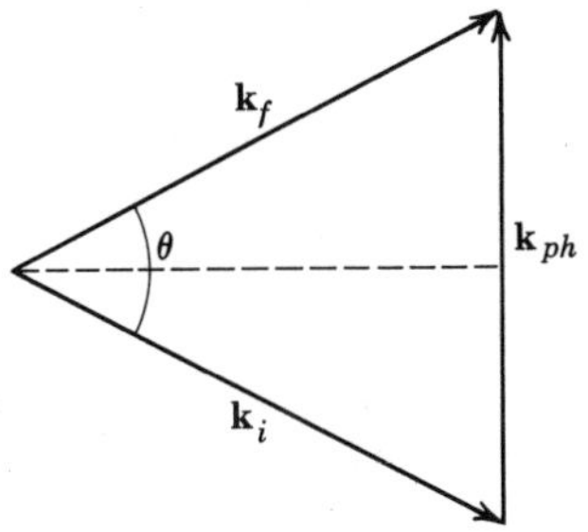

Fig. 12.1. Momentum diagram for the scattering of an electron by a phonon.

If we now come back to the matrix element M, we have

$$M \propto \int g_f^*(\boldsymbol{\xi}\cdots)W(\mathbf{r}, \boldsymbol{\xi})g_i(\boldsymbol{\xi}\cdots)\, d\boldsymbol{\xi}_a\cdots d^3\mathbf{r} \tag{12.9}$$

Now, if W varies slowly with position, we can replace $W(\mathbf{r}, \boldsymbol{\xi})$ by its average over r, which we call $\langle W(\boldsymbol{\xi})\rangle_r$, and we are left with an integral over the $\boldsymbol{\xi}$'s only:

$$M \propto \int g_f^*(\boldsymbol{\xi}_{k'})\langle W(\boldsymbol{\xi})\rangle_r g_i(\boldsymbol{\xi}_k)\, d\boldsymbol{\xi}\cdots \tag{12.10}$$

The quantity $\langle W(\boldsymbol{\xi})\rangle_r$ represents the Hamiltonian of the phonon wave. If we disregard the terms of order higher than the second in the potential of the lattice, we are facing the same problem as in Sec. 4 concerning the nuclear wave equation.

We remember that the matrix elements between two states of the lattice were calculated with the help of the creation and destruction operators (Eqs. 4.12 and 4.13). Here, the same method permits the evaluation of M, in terms of the integrals

$$\int g_f^* a_{f-i}^+ g_i\, d\boldsymbol{\xi} = \frac{1}{\sqrt{2M_a\omega_{\mathbf{k}}^2}}\sqrt{r_{\mathbf{k}}+1} \tag{12.11a}$$

and

$$\int g_f^* a_{f-i} g_i\, d\boldsymbol{\xi} = \frac{1}{\sqrt{2M_a\omega_{\mathbf{k}}^2}}\sqrt{r_{\mathbf{k}}} \tag{12.11b}$$

The number $r_{\mathbf{k}}$ of lattice modes of energy $\hbar\omega_{\mathbf{k}}$ is given by Planck's formula

$$r_{\mathbf{k}} = \left[\exp\left(\frac{\hbar\omega_{\mathbf{k}}}{kT_L}\right) - 1\right]^{-1} \tag{12.12}$$

where T_L is the temperature of the lattice, and

$$\hbar\omega_{\mathbf{k}} = c\hbar|\mathbf{k}_{ph}| = 2c\hbar|\mathbf{k}_i| \sin\theta/2 \tag{12.13}$$

by virtue of Eq. 12.8.

Consequently, $r_{\mathbf{k}}$ can be rewritten as follows:

$$r_{\mathbf{k}} = \left[\exp\left(\frac{2c\hbar|\mathbf{k}_i|\sin\theta/2}{kT_L}\right) - 1\right]^{-1} \tag{12.14}$$

In insulators, semiconductors, and also metals much above the Debye temperature, we have

$$\frac{c\hbar|\mathbf{k}|}{kT_L} << 1$$

so that Eq. 12.14 can be written as

$$r_{\mathbf{k}} \simeq r_{\mathbf{k}} + 1 \simeq \frac{kT_L}{2c\hbar|\mathbf{k}| \sin \theta/2}$$

and $|M|^2$ is proportional to $r_{\mathbf{k}}/\omega_{\mathbf{k}}^2$.

Our next task is to calculate $\rho(E_{\mathbf{v}})$.

Calculation of the Electronic Density $\rho(E_{\mathbf{v}})$

The calculation of $\rho(E_{\mathbf{v}})$ is easy if we assume that the effective mass is isotropic around the minimum of the conduction band. In this case,

$$E_{\mathbf{v}} = \tfrac{1}{2}m^*v^2$$

hence,

$$\rho(E_{\mathbf{v}}) = \frac{dE_{\mathbf{v}}}{dv} = m^*v$$

Our final task is to calculate $R(E_{\mathbf{k}_{ph}})$.

Calculation of the Phonon Density $R(E_{\mathbf{k}_{ph}})$

This is also a familiar problem. The concentration, in a box, of modes with wave vectors smaller than a given vector **k** is

$$R(E_{\mathbf{k}_{ph}})\, dE_{\mathbf{k}_{ph}} = \tfrac{1}{8}(4\pi k_{ph}^2\, dk_{ph})$$

Since

$$|\mathbf{k}_{ph}| = \frac{\omega_{\mathbf{k}}}{c} = \frac{E_{\mathbf{k}_{ph}}}{\hbar c}$$

then

$$R(E_{\mathbf{k}_{ph}})\, dE_{\mathbf{k}_{ph}} = \frac{\pi}{2} \frac{\omega_{\mathbf{k}}^2}{\hbar c^3}\, dE_{\mathbf{k}_{ph}}$$

or

$$R(E_{\mathbf{k}_{ph}}) = \frac{\pi}{2\hbar c^3}\, \omega_{\mathbf{k}}^2$$

From the expressions given for $|M^2|$, $\rho(E_{\mathbf{v}})$, and $R(E_{\mathbf{k}_{ph}})$, it results that $P(\mathbf{v}, \mathbf{v}')$ is proportional to $m^*r_{\mathbf{k}}v$, that is, to the quantity

$$\frac{T_L m^* v}{k \sin \theta/2}$$

and since $k = (m^*/\hbar)v$, $P(\mathbf{v}, \mathbf{v}')$ is finally proportional to $T_L/(\sin \theta/2)$.

13. Mean Free Time for Phonon Scattering

Let us consider an electron with an instantaneous velocity $\mathbf{v}$ at $t = 0$, and let the initial position of the electron be the origin ($\mathbf{r}(0) = 0$). If we define the average velocity of that electron at t as

$$\langle \mathbf{v} \rangle_t = \frac{\mathbf{r}(t)}{t} = \frac{1}{t} \int_0^t \mathbf{v}(t)\, dt \tag{13.1}$$

we see that

$$\langle \mathbf{v} \rangle_t \to 0 \quad \text{as} \quad t \to \infty$$

if no field is applied.

The purpose of this section is to show that $\langle \mathbf{v} \rangle_t$ decreases exponentially with t, so that we can define a mean free time τ such that

$$\langle \mathbf{v} \rangle_t = \langle \mathbf{v} \rangle_0 \exp\left(-\frac{t}{\tau}\right) = \mathbf{v} \exp\left(-\frac{t}{\tau}\right) \tag{13.2}$$

We have seen previously that an elastic collision with the lattice changes only the direction of $\mathbf{v}$, so that

$$|\mathbf{v}'| = |\mathbf{v}| \tag{13.3}$$

Therefore, if we assume that $\mathbf{v}'$ can take any orientation with respect to $\mathbf{v}$, we can write

$$\left\langle \frac{d\mathbf{v}}{dt} \right\rangle_t = \frac{d}{dt} \langle \mathbf{v} \rangle_t = \int (\mathbf{v}' - \mathbf{v}) P(\mathbf{v}, \mathbf{v}')\, d^3\mathbf{v} \tag{13.4}$$

where $\mathbf{v}'$ is some radius vector of the sphere S of radius v in velocity space (Fig. 13.1) and $P(\mathbf{v}, \mathbf{v}')$ the probability per unit time that an electron of velocity $\mathbf{v}$ will acquire a velocity $\mathbf{v}'$ within the volume element $d^3\mathbf{v}$ of velocity space.

In view of the relation in Eq. 13.3 and of the symmetry of revolution around $\mathbf{v}$, we can make the following two statements:

a. $d^3\mathbf{v}$ reduces to the area of an elementary ring on the sphere S:

$$d^3\mathbf{v} = 2\pi v \sin\theta\, d\theta$$

where θ is the angle $(\mathbf{v}, \mathbf{v}')$.

b.

$$\mathbf{v}' - \mathbf{v} = \mathbf{v}(\cos\theta - 1) + \mathbf{v}_\perp \sin\theta$$

where $\mathbf{v}_\perp$ is a vector of magnitude v, perpendicular to $\mathbf{v}$ in the plane $\mathbf{v}, \mathbf{v}'$.

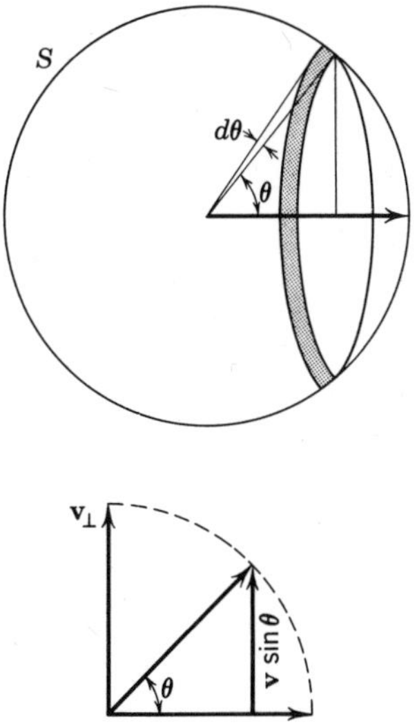

Fig. 13.1. Solid angle in velocity space.

Finally, Eq. 13.4 can be written

$$\frac{d}{dt}\langle\mathbf{v}\rangle_t = -2\pi \int_0^\pi [(1 - \cos\theta)\mathbf{v} - \sin\theta\, \mathbf{v}_\perp] P(\mathbf{v}, \mathbf{v}') v \sin\theta\, d\theta$$

In view of the symmetry of revolution around $\mathbf{v}$, $\langle\mathbf{v}_\perp\rangle$ vanishes much faster than $\langle\mathbf{v}\rangle$, so that, for any time t not too small, the above expression reduces to

$$\frac{d}{dt}\langle\mathbf{v}\rangle_t = -2\pi \int_0^\pi (1 - \cos\theta)\mathbf{v} P(\mathbf{v}, \mathbf{v}') v \sin\theta\, d\theta \qquad (13.5)$$

Furthermore, under the same assumption as above,

$$\int_0^\pi (1 - \cos\theta)\mathbf{v} P(\mathbf{v}, \mathbf{v}') v \sin\theta\, d\theta = \langle\mathbf{v}\rangle_t \int_0^\pi (1 - \cos\theta) P(\mathbf{v}, \mathbf{v}') v \sin\theta\, d\theta$$

so that Eq. 13.5 becomes

$$\frac{d}{dt}\langle\mathbf{v}\rangle_t = -2\pi\langle\mathbf{v}\rangle_t \int_0^\pi (1 - \cos\theta) P(\mathbf{v}, \mathbf{v}') v \sin\theta\, d\theta \qquad (13.6)$$

We have seen that $P(\mathbf{v}, \mathbf{v}')$ is a function of θ and of some parameters of the lattice. For instance, in the simple case where T_L is much larger than the Debye temperature Θ,

$$P(\mathbf{v}, \mathbf{v}') \propto \frac{T_L}{\sin \theta/2}$$

Consequently, the integral is a constant parameter having the dimension of t^{-1}, and we can define a mean free time τ such that

$$\frac{1}{\tau} = 2\pi \int_0^\pi (1 - \cos \theta) P(\mathbf{v}, \mathbf{v}') v \sin \theta \, d\theta \tag{13.7}$$

With this definition, we have

$$\frac{d}{dt} \langle \mathbf{v} \rangle_t = -\frac{\langle \mathbf{v} \rangle_t}{\tau}$$

or

$$\langle v \rangle_t = \langle v \rangle_0 \exp \left(-\frac{t}{\tau} \right)$$

as predicted.

The mean free time defined above can be introduced to define a "mean free path" by the relation

$$\langle \mathbf{l} \rangle = \int_0^\infty \langle \mathbf{v} \rangle \, dt = \langle \mathbf{v} \rangle_0 \int_0^\infty \exp \left(-\frac{t}{\tau} \right) dt$$

or

$$\langle \mathbf{l} \rangle = \tau \langle \mathbf{v} \rangle_0$$

Normal Temperature Dependence

Let us now return to our case $T >> \Theta$; in this case, Eq. 13.7 becomes

$$\frac{1}{\tau} \propto \int_0^\pi (1 - \cos \theta) \frac{T_L}{\sin \theta/2} v \sin \theta \, d\theta$$

showing that

$$\tau \propto \frac{1}{v T_L}$$

Now, if we want to find the average mean time $\langle \tau \rangle$ for all electrons, τ must be weighted with the distribution function $v^4 \exp(-m^* v^2/2kT_e)$, introduced in Sec. 11, where T_e = temperature of the electron gas:

$$\langle \tau \rangle = \frac{1}{T_L} \langle v^{-1} \rangle = \frac{\int_0^\infty v^4 \exp(-m^* v^2/2kT_e) v^{-1} \, dv}{\int_0^\infty v^4 \exp(-m^* v^2/2kT_e) \, dv}$$

The calculation of the integrals is straightforward and shows that $\langle\tau\rangle$ varies as $T_L^{-1}T_e^{-1/2}$.

Under a weak applied field, the electrons can be considered in equilibrium with the lattice, so that $T_e \approx T_L$ and

$$\langle\tau\rangle \propto T_L^{-3/2} \tag{13.8}$$

This result is based on the assumption that only longitudinal phonons produce the scattering. The introduction of the transverse phonons would change only the constant of proportionality.

Since the conductivity is proportional to $\langle\tau\rangle$, we have established that the ratio σ/n (conductivity per unit concentration of carriers) varies as $T^{-3/2}$ in semiconductors under low electric field. This result has been verified with intrinsic semiconductors in which a fixed concentration of free carriers is produced by irradiation.

In the case of metals, $\langle v^{-1}\rangle$ is constant, because the averaging function from Fermi-Dirac statistics is essentially a delta function, centered at the Fermi energy. Consequently, $\langle\tau\rangle$ is inversely proportional to T_L, except at low temperatures, and we find the well-known result that the resistivity of a metal is proportional to its absolute temperature.

Phonon Drag

We have seen that the interaction between electrons and phonons is stronger when both have comparable wave vectors. In other words, more energy is transferred by the electrons to those phonons for which $k_{ph} \approx kT_e/\hbar c$. If the relaxation time between phonons of different energies is much longer than that between electrons and phonons, the net result is a displacement of the whole energy spectrum of the phonons. The situation, known as the "phonon drag," manifests itself in semiconductors by an unusually high thermoelectric power at low temperature, and in metals by a T^5 dependence of the resistivity at low temperature.

The matrix element M, which contributes to the temperature dependence of the mean free time, was calculated under the assumption that the energy of the phonons is negligible compared to that of the electrons (elastic scattering). However, this does not hold at very low temperatures, where the electron energy becomes comparable to the zero-point energy of the lattice.

Furthermore, in the case of metals, the Brillouin zone is almost filled, so that only a thin region of states is available between the Fermi surface and the Brillouin zone edges. Consequently, electrons can interact only with phonons of small wave number. If $T > \Theta$, the average value $\langle|\mathbf{k}_{ph}|\rangle$ of the phonon wave number is roughly independent of T, and is given by

$$k\Theta \simeq c\langle|\mathbf{k}_{ph}|\rangle$$

When $T < \Theta$, $\langle|\mathbf{k}_{ph}|\rangle$ decreases with T (more phonons have low energy), and the temperature dependence of the scattering is enhanced.

Actually, a T^5 dependence of the resistivity has been observed, around a few degrees Kelvin, for some nonsuperconductive metals. However, the resistivity of such metals never goes to zero at vanishing temperature, but instead tends towards a finite limit relevant to the imperfections and impurities of the material.

14. Nonintrinsic Scattering Mechanism

Scattering by Ionized Impurities

Suppose, for instance, that a covalent semiconductor (germanium or silicon) contains pentavalent impurity atoms such as phosphorus. One of the valence electrons of these atoms cannot participate in the bonds, and the system formed by the positively charged phosphorus core and the extra electron can be considered as a hydrogen atom. However, the energy levels of this system are considerably reduced with respect to those of the hydrogen atom by the factor $\kappa^2 m/m^*$, where κ is the relative dielectric constant and m^* the electronic effective mass of the host crystal. In particular, the new ionization energy is often of the order 0.01 ev, which is smaller than the thermal energy kT at room temperature (0.025 ev). In such cases, most of the impurity atoms are ionized at room temperature, and act as point charges that can scatter conduction electrons.

Since electrons are much lighter than atoms, the scattering centers can be regarded as fixed, and the collisions change only the direction of the electron velocity but not its magnitude. In other words, the situation is comparable to that encountered in the scattering by phonons (Fig. 12.1).

In order to attempt a quantitative discussion of the problem, we shall first assume that the potential created by the centers does not affect the distribution of electrons around them. Under this circumstance, the potential energy of an electron at the distance r from such a center is $U = -e^2/r$, and the matrix element of the perturbation U between the initial state and the final state is proportional to the integral $\int \exp[i(\mathbf{k}' - \mathbf{k})\cdot\mathbf{r}]U\, d^3\mathbf{r}$, which is merely the Fourier transform of U. Since the volume element contains $r^2\, dr$, M is proportional to $e^2 \int \exp[i(\mathbf{k}' - \mathbf{k})\cdot\mathbf{r}]r\, dr$, which varies as $e^2|\mathbf{k}' - \mathbf{k}|^{-2}$, that is, as $(v \sin\theta/2)^{-2}$. The probability $P(\mathbf{v}, \theta)\, d\Omega$ that an electron of speed $\mathbf{v}$ undergoes a transition to a state of speed $\mathbf{v}'$ in the solid angle element $d\Omega$ around θ is proportional to M^2 and, hence, varies as $(v \sin\theta/2)^{-4}$. The probability for a transition to any state $\mathbf{v}' \neq \mathbf{v}$ is obtained by averaging $P(v, \theta)\, d\Omega$ with the appropriate weighting factor over all orientations. From the study of Sec. 4 and the assumed symmetry of revolution around the direction of the initial velocity $\mathbf{v}$, it results that the weighting factor is proportional to the projection of $\Delta\mathbf{v} = \mathbf{v}' - \mathbf{v}$ on $\mathbf{v}$, that is, $v(1 - \cos\theta)$.

Finally, the reciprocal mean free time for a velocity v is such that

$$\tau^{-1} \propto v^{-3} \int_0^\pi (1 - \cos\theta)(\sin\theta/2)^{-4}\, d(\cos\theta) = v^{-3} \int_0^\pi (1 - \cos\theta)^{-1}\, d(\cos\theta) \quad (14.1)$$

The integral in Eq. 14.1 diverges, and this divergence is expected, since an unshielded Coulomb field still introduces a small amount of scattering at infinite distance. In other words, the total cross section of a charged center for an electron (when both are *in vacuo*) is infinite.

In actual solids, however, the conduction electrons build up a space charge around the centers, thereby shielding their electric field, just as does the "Debye" atmosphere surrounding the ions in an electrolyte.

Due to this space charge, the field around the centers drops much more rapidly than if it were an unshielded Coulomb field, and the scattering cross section assumes a finite value. In fact, the shielding can be considered as changing the lower limit of the integral in Eq. 14.1 from 0 to $\theta_0 \neq 0$.

Before discussing quantitatively the effect of shielding, we can find an approximation for the temperature dependence of the average mean free time. We have also seen in Eq. 14.1 that $\tau(v)$ is proportional to v^3. Hence, for a semiconductor in which electrons obey a Maxwellian distribution, the average mean free time is obtained with the same weighting function as earlier and takes the form

$$\langle\tau(v)\rangle = \frac{\int_0^\infty v^7 \exp(-m^*v^2/2kT)\,dv}{\int_0^\infty v^4 \exp(-m^*v^2/2kT)\,dv} \tag{14.2}$$

which is proportional to $T^{3/2}$.

For metals, where electrons obey the Fermi-Dirac statistics, $\langle\tau(v)\rangle$ is independent of temperature.

We shall now discuss quantitatively the effect of the shielding space charge.

The electronic current density at any point in a solid is

$$\mathbf{j} = \mu e n \mathscr{E} - De\,\nabla n \tag{14.3}$$

(a similar expression can be written for holes). In this relation,

μ = the mobility of the electrons

n = their concentration

D = their diffusion constant

$\mathscr{E}$ = the electric field

If no external field is applied, $\mathscr{E}$ results from the presence of the ionized impurities and of the space charge. The charge-density distribution at equilibrium is the superposition of a density ρ_1 corresponding to the Coulomb potential U of the centers, given by Poisson's equation

$$\rho_1 = \kappa\epsilon_0\,\nabla^2 U \tag{14.4}$$

and of a density ρ_2 due to the space charge itself, which is merely

$$\rho_2 = e\,\delta n \tag{14.5}$$

where δn is the excess concentration, with respect to the averaged, uniform concentration n_0.

Finally, the total charge density is

$$\rho = \rho_1 + \rho_2 = \kappa\epsilon_0\,\nabla^2 U + e\,\delta n \tag{14.6}$$

and Poisson's equation for the entire system is

$$\nabla\cdot\mathscr{E} = \frac{\rho}{\kappa\epsilon_0} = \nabla^2 U + \frac{e}{\kappa\epsilon_0}\,\delta n \tag{14.7}$$

Furthermore, if we use Einstein's relation

$$\frac{\mu}{D} = \frac{e}{kT} \tag{14.8}$$

in Eq. 14.3, in which $\mathbf{j} = 0$ for the steady state, we get

$$\frac{\mu}{D} n\mathscr{E} = \frac{e}{kT} n\mathscr{E} = \nabla n \tag{14.9}$$

or, since $n = n_0 + \delta n \simeq n_0$,

$$\mathscr{E} \simeq \frac{kT}{n_0 e} \nabla \, \delta n \tag{14.10}$$

Let us now introduce a potential function ϕ such that the field $\mathscr{E}$ derives from ϕ:

$$\mathscr{E} = -\nabla\phi \tag{14.11}$$

The new potential ϕ is called the "shielded" potential, as the field $\mathscr{E}$, shielded by the space charge, derives from it.

Introducing this shielded potential in Eq. 14.10, we get

$$-\nabla^2\phi = \frac{kT}{en_0} \nabla^2 \, \delta n \tag{14.12}$$

and, by taking the Fourier transform of both sides, we obtain

$$-p^2 F(\phi) = \frac{kT}{en_0} p^2 F(\delta n) \tag{14.13}$$

where $F(\phi)$, the Fourier transform of $\phi(r)$, is given by

$$F(b) = \frac{1}{\sqrt{2\pi}} \int \exp(-ipr)\phi(r) \, dr \tag{14.14}$$

Doing exactly the same with Eq. 14.7, we get

$$-\nabla^2\phi = \nabla^2 U + \frac{e}{\epsilon_0 \kappa} \delta n \tag{14.15}$$

which, by Fourier transformation, gives

$$-p^2 F(\phi) = p^2 F(U) + \frac{e}{\kappa\epsilon_0} F(\delta n) \tag{14.16}$$

We can now eliminate $F(\delta n)$ between Eqs. 14.15 and 14.16. This gives

$$-p^2 F(\phi) = p^2 F(U) + \frac{e^2 n_0}{\kappa\epsilon_0 kT} F(\phi)$$

or

$$F(\phi) = -\frac{p^2}{p^2 + \dfrac{e^2 n_0}{\kappa\epsilon_0 kT}} F(U) \tag{14.17}$$

If we now take the inverse transform of both sides, we see that

$$\phi = U \exp\left[-\left(\frac{e^2 n_0}{\kappa\epsilon_0 kT}\right)^{1/2} r\right] \tag{14.18}$$

or

$$\phi = \frac{e^2}{r}\exp\left(-\frac{r}{L}\right) \tag{14.19}$$

where

$$L = \left(\frac{\kappa\epsilon_0 kT}{e^2 n_0}\right)^{1/2} \tag{14.20}$$

is the "Debye length," that is, the distance at which the scattering field is $1/e$ of what it would be without shielding.

The matrix element of the shielded potential ϕ between the initial and final states for a given v has been calculated by Brooks and Herring, who found that

$$\frac{1}{\tau(v)} = \frac{\pi}{m^{*2}v^3}\frac{e^4 N}{\kappa^2\epsilon_0{}^2}\left[\ln(1+b) - \frac{b}{1+b}\right] \tag{14.21}$$

where

$$b = \frac{4\pi m^{*2}\kappa\epsilon_0 kT}{e^2\hbar^2 n_0}v^2 = \left(\frac{m^* L}{\hbar\sqrt{\pi}}\right)^2 v^2 \tag{14.22}$$

Since b is usually much larger than unity, the term in the brackets reduces approximately to $\ln b$, so that

$$\frac{1}{\tau(v)} \simeq \frac{2\pi}{m^* v^3}\frac{e^4 N}{\kappa^2\epsilon_0{}^2}\ln\left(\frac{m^* v}{\hbar\sqrt{\pi}}L\right) \tag{14.23}$$

The last step is to average the reciprocal mean free time given by Eq. 14.23 by means of the weighting function $v^4 \exp(-m^* v^2/2kT)$. Unfortunately, the logarithmic term now creates mathematical difficulties, but a good approximation can be obtained by taking the logarithm outside the integral and replacing its argument by its value assumed when the rest of the integrant is maximum.

The resulting mobility, calculated by Brooks and Herring, is

$$\mu = \frac{2^{7/2}(\epsilon_0\kappa)^2(kT)^{3/2}}{\pi^{3/2}e^3 m^{*1/2}N}\left[\ln\frac{6m^*(\kappa\epsilon_0)(kT)^2}{\pi e^2\hbar^2 n_0(2 - n_0/N)}\right]^{-1} \tag{14.24}$$

In a semiconductor at room temperature, practically all the impurities are ionized ($n_0 \simeq N$), so that Eq. 14.24 reduces to

$$\mu = \frac{2^{7/2}(\epsilon_0\kappa)^2(kT)^{3/2}}{\pi^{3/2}e^3 m^{*1/2}N}\left[\ln\frac{6m^*\kappa\epsilon_0(kT)^2}{\pi e^2\hbar^2 N}\right]^{-1} \tag{14.25}$$

We can see that the $T^{3/2}$ temperature-dependence factor obtained in the approximate treatment neglecting shielding is only slightly modified by the temperature dependence of the logarithmic term.

In metals, the high concentration of free electrons (about one per atom) manifests itself in two ways. The shielding is much more complete than in the semiconductors, and the electrons obey the Fermi-Dirac statistics. It can be shown in this case that the average mean free path due to impurity scattering is independent of temperature.

Neutral-Impurity Scattering

We have seen previously that a charged center surrounds itself with a shielding space charge that is characterized by its Debye length, given by Eq. 14.20. Hence, we can distinguish between an ionized and a neutral impurity by comparing the first Bohr radius $r_1 = \kappa(m/m^*)a_0$ and the Debye length L.

If $r_1 \geq L$, the extra electron is only very weakly bound to the core, and the impurity can be considered as ionized.

If $r_1 \leq L$, the electron is bound to the core by a nearly Coulomb field, and the impurity can be considered as neutral.

Although the mean free time due to scattering by neutral and ionized impurities is of the same order of magnitude, the processes are notably different. In the present case, the scattering of low-energy electrons involves the exchange of a bound electron with the incoming electron.

Calculations show that the inverse mobility arising from neutral-impurity scattering is of the form

$$\mu^{-1} \propto T^{1/2} f(T) \tag{14.26}$$

where $f(T)$ is a function that decreases gradually from 1 to 0 when T increases from 0 to ∞. In practice, neutral-impurity scattering is expected to be observed only at very low temperature, since otherwise all impurities are ionized.

Scattering by Dislocations

The scattering arises from the strain field around the dislocations. The perturbing field due to a linear edge dislocation can be written in the form

$$U = A \sin \theta / r \tag{14.27}$$

where A is a parameter containing the slip distance and the Poisson ratio of the material, r is the distance from the axis of the dislocation, and θ is the angle measured from the slip direction.

The same technique as that described for the scattering by ionized impurities can be used to show that the average reciprocal mean free time of electrons scattered by edge dislocations is proportional to the reciprocal temperature. In other words, the corresponding mobility is proportional to the absolute temperature.

The density of dislocations must be high to produce a scattering mechanism that competes with the others, and it is only at very low temperatures that the scattering by dislocations can eventually be observed.

Electron-Electron Scattering

The scattering of electrons by electrons tends to randomize the momentum distribution among the electrons. Consequently, it cannot affect the mobility when the momentum gained from the field between collisions with impurities is very small. However, it may become effective under high applied fields, for the following reason. The study of the scattering by impurities shows that the scattering cross section for electrons of high energy is much smaller than for electrons of low energy. Hence, the mean free path of high-energy electrons for impurity scattering may become larger than for electron scattering, and electron-electron scattering can redistribute among the entire electron distribution the momentum gained from the field between collisions, thus reducing the mobility. This point will be discussed further in the next section.

Comparative Study of the Various Scattering Processes under Low Applied Field

By far the predominant two scattering mechanisms in the usual temperature range are the lattice scattering and the ionized-impurity scattering.

If both processes are present simultaneously, we can assume that the transition probabilities are additive. Since the transition probability is proportional to the reciprocal mean free time, which is itself proportional to the reciprocal mobility, by virtue of Eq. 11.12, we can write

$$\mu^{-1} = \mu^{-1}{}_{\text{phonons}} + \mu^{-1}{}_{\text{impurities}} \tag{14.28}$$

In semiconductors, Eq. 14.28 takes the form

$$\mu^{-1} = AT^{3/2} + BT^{-3/2} \tag{14.29}$$

where A and B are coefficients independent of the temperature. Equation 14.29 predicts that the mobility passes through a maximum at the temperature for which both scattering mechanisms are equally effective, and this prediction is confirmed by experiment.

In metals, Eq. 14.28 takes the form

$$\mu^{-1} = CT + D \tag{14.30}$$

where the constant D refers to the residual resistance at vanishing temperatures.

Comparison with Experiments

The mobility of the carriers and its temperature dependence cannot usually be obtained directly by conductivity measurements, since the conductivity σ involves the number of contributing carriers. For instance, if only electrons (or holes) are involved, we have

$$\mu = \frac{\sigma}{en}$$

The carrier concentration σ is usually a function of temperature, except for metals and also for some semiconductors in a narrow temperature range between the intrinsic range where carriers are thermally excited across the gap and that in which the impurities are not completely ionized.

Since the Hall effect gives direct information concerning the carrier concentration, combined measurement of the conductivity and the Hall coefficient R are required to find μ. In particular, in the case of a nondegenerate semiconductor in the extrinsic range and containing only one type of impurity, μ is proportional to σR.

Combined measurements of conductivity and Hall coefficients were used in the late forties at the Bell Telephone Laboratories to find the mobility in germanium crystals doped with various concentrations of arsenic. Typical results are shown in Fig. 14.1, in which the concentration of impurities increases from sample A to sample D. In sample A, which is almost intrinsic, the mobility follows very closely the $T^{-3/2}$ law of lattice scattering. In samples B, C, and D, impurity scattering, which sets in at increasing temperatures, lowers the mobility and reverses its temperature coefficient. It should be noticed on curve D that the mobility tends to rise slightly with decreasing temperature below 30°K. This might arise from the fact that the temperature dependence of the logarithmic factor in Eq. 14.25 prevails over the $T^{3/2}$ factor when T is very small, since we have in this case

$$\frac{T^{3/2}}{\ln(1 + \alpha T^2)} \sim \frac{T^{3/2}}{\alpha T^2} = \alpha^{-1} T^{-1/2}$$

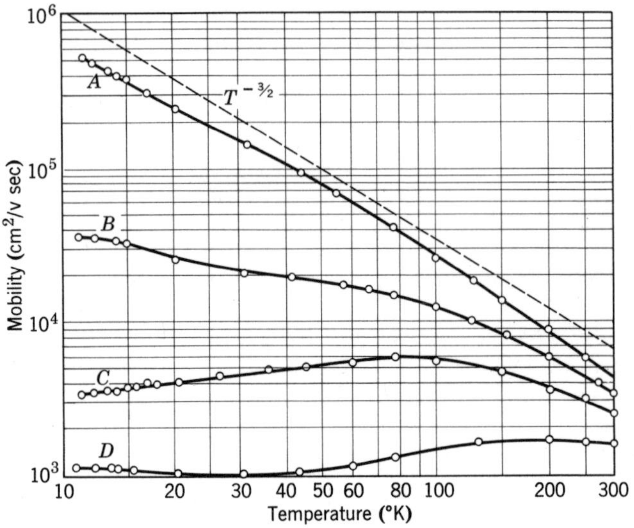

Fig. 14.1. Mobility versus temperature for arsenic-doped germanium samples. Impurity concentration increases from sample *A* to sample *D*. The data were taken by Debye at Bell Laboratories. The dashed line represents a $-\frac{3}{2}$ slope characteristic of lattice scattering.

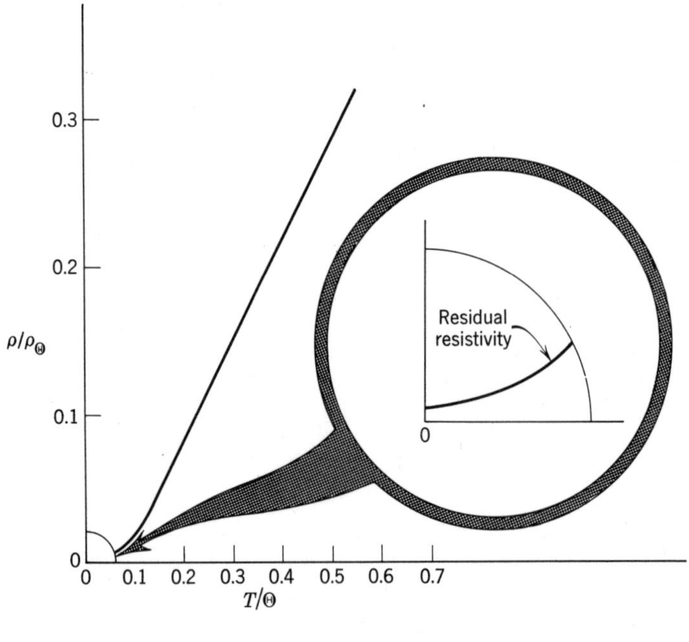

Fig. 14.2. Typical temperature dependence of the resistivity in nonsuperconducting metals.

In metals, the number of free electrons per atom is of the order of unity, and is independent of temperature. Hence, the mobility is proportional to the conductivity. Figure 14.2 represents the resistivity of sodium metal as a function of the absolute temperature. As predicted, the resistivity varies linearly with the temperature, except at very low temperature, where the phonon drag enters into play. The extrapolated value of the resistivity for 0°K is finite, and results from the presence of impurities.

15. High-Field Conduction Phenomena

Until now, the conductivity phenomena have been discussed under the assumption that the applied field was very small. By this, we meant that the average kinetic energy gained by the carriers from the field between collisions was negligible compared to their thermal energy, and that the phonon system could reach a steady-state equilibrium. As a consequence of these assumptions, the mean free time and mobilities obtained in the previous sections were independent of the field, and this prediction of Ohm's law was verified experimentally for germanium up to field values of the order of 100 v/cm.

If the applied field is high, most of the assumptions on which the low-field theory was based now break down. In particular, the nature of the scattering processes is profoundly modified. The scattering by ionized impurities becomes less effective, the collisions with phonons become appreciably inelastic, and electron-electron scattering becomes an important factor in limiting the transfer of energy. Consequently, the energy distribution of the electron gas cannot be purely Maxwellian. Besides, we must keep in mind that the effective mass of the carriers depends on the applied field, since it is related to the curvature of the conduction band at the point of **k**-space which represents the average drift momentum of the electron gas.

Measurements of the mobility under high-field conditions require the use of very short voltage pulses, since a high direct current would result in gradual heating of the sample up to thermal breakdown. The first reliable measurements of the high-field mobility in germanium were made at the Bell Telephone Laboratories in the early fifties by E. J. Ryder, using pulses of microsecond duration. These measurements showed that the mobility drops when the field becomes larger than some value that depends on the temperature and the impurity content of the sample.

Shockley has discussed Ryder's results, under the assumption that the electrons obey a Maxwellian distribution of the form

$$f(-\mathbf{v}) = A \exp\left(\frac{m^* v^2}{2kT_e}\right) \tag{15.1}$$

where T_e, the temperature of the electron gas, is larger than that of the lattice and behaves as an increasing function of the applied field. Qualitative agreement with Ryder's results was obtained by Shockley, but further measurements have shown that Shockley's theory is only approximate.

We shall now summarize a theory developed by J. Bok in his thesis (University of Paris, 1959, unpublished), and which we might call the approximation of the electronic temperature.

If the applied field is high, the interelectronic collisions change the distribution function of the electrons and introduce a nonlinear term in Boltzmann's equation.

If one assumes that the modified distribution function depends upon two parameters, two independent relations involving the distribution function are sufficient to determine the parameters.

Under the assumption that the energy distribution results exclusively from interelectronic collisions, the distribution function would be of the form

$$f(\mathbf{v}) = A \exp\left[-\frac{m^*(\mathbf{v} - \mathbf{v}_0)^2}{2kT_e}\right] \tag{15.2}$$

where A is a normalization constant, $\mathbf{v}_0$ the drift velocity of the electrons caused by the applied field, and T_e the temperature of the electron gas.

In fact, the interelectronic collisions do not play an exclusive role, since the phonons also contribute to the scattering. Nevertheless, one can attempt to represent the energy distribution by a relation of the form of Eq. 15.2 and to obtain $\mathbf{v}_0$ and T_e as a function of $\mathcal{E}$ from the equations for the conservation of energy and total momentum in the steady state. One can show that this procedure is legitimate at room temperature when the carrier concentration exceeds 10^{14} cm^{-3}, corresponding to relatively pure germanium.

As we shall see later, kT_e is usually much larger than $\frac{1}{2}m\mathbf{v}_0^2$, so that we can use for $f(\mathbf{v})$ the expansion

$$f(\mathbf{v}, \mathbf{v}_0) = f(\mathbf{v}, 0) + \mathbf{v}_0 \cdot \frac{m^*}{kT_e} \mathbf{v} f(\mathbf{v}, 0) \tag{15.3}$$

with

$$f(\mathbf{v}, 0) = A \exp\left(-\frac{m^* v^2}{2kT_e}\right) \tag{15.4}$$

Conservation of Energy

In the steady state,

$$\left(\frac{dE}{dt}\right)_{\text{field}} = \left(\frac{dE}{dt}\right)_{\text{collisions}} \tag{15.5}$$

that is,

$$ne\mathbf{v}_0 \cdot \mathscr{E} = \frac{m^*}{2} \iint (v^2 - v'^2) P(\mathbf{v}, \mathbf{v}') f(\mathbf{v})\, d^3\mathbf{v}\, d^3\mathbf{v}' \tag{15.6}$$

where $\mathbf{v}$ and $\mathbf{v}'$ are the velocities before and after collision, $P(\mathbf{v}, \mathbf{v}')$ the probability of a transition $\mathbf{v} \to \mathbf{v}'$, and $f(\mathbf{v})$ the distribution function. By replacing $f(\mathbf{v})$ by the first term of its expansion given in Eq. 15.3, Eq. 15.4 takes the form

$$ne\mathbf{v}_0 \cdot \mathscr{E} = F(T_e) \tag{15.7}$$

where the function $F(T_e)$ depends only on T_e and on the form of the probability $P(\mathbf{v}, \mathbf{v}')$.

Conservation of Momentum

Under the assumption of an isotropic effective mass, the conservation of momentum takes the form

$$\left(\frac{d\mathbf{v}}{dt}\right)_{\text{field}} = \left(\frac{d\mathbf{v}}{dt}\right)_{\text{collisions}} \tag{15.8}$$

that is,

$$-ne\mathscr{E} = m^* \iint (\mathbf{v} - \mathbf{v}') P(\mathbf{v}, \mathbf{v}') f(\mathbf{v})\, d^3\mathbf{v}\, d^3\mathbf{v}' \tag{15.9}$$

Here, the contribution of the first term of Eq. 15.3 averages to zero, since there is no current when $\mathbf{v}_0 = 0$, that is, in the absence of an applied field. When the second term of the expansion is used for $f(\mathbf{v})$, Eq. 15.9 takes the form

$$ne\mathcal{E} = \mathbf{v}_0 G(T_e) \tag{15.10}$$

where the function $G(T_e)$ depends only on T_e and on the form of the probability $P(\mathbf{v}, \mathbf{v}')$.

Equations 15.7 and 15.10 constitute a system from which $\mathbf{v}_0$ and T_e can be derived as functions of $\mathcal{E}$.

The functions $F(T_e)$ and $G(T_e)$ can be calculated in a few simple cases, namely, when the phonon distribution takes a simple form (that is, when the energy $\hbar\omega_{ph}$ of a phonon is either very small or very large compared to kT), and under the assumption that the energy of the phonon is small compared to that of the electron, so that the densities of states $\rho(E)$ and $\rho(E')$ are the same. Under these conditions, one finds for the scattering by acoustical phonons

$$F(T_e) \propto \begin{cases} T_e \text{ for low temperatures} \\ TT_e^{1/2} \text{ for high temperatures} \end{cases}$$

$$G(T_e) \propto T_e^{3/2}$$

This approximation of $G(T_e)$ is evidently wrong for low applied field, since $G(T_e)$ should vanish with the difference $(T_e - T)$. The difficulty can be obviated by using the empirically corrected relation $G(T_e) \propto (T_e - T)T_e^{1/2}$, which vanishes for $T_e = T$ and varies as $T_e^{3/2}$ for $T_e/T >> 1$. Similar calculations have been performed for the scattering by optical phonons and ionized impurities.

The field and temperature variations of the current resulting from the above values of $F(T_e)$ and $G(T_e)$ are in reasonable agreement with the experimental results.

An important justification of the hot-electron approximations lies in the observation, made by Bok and verified by several workers, that the electrons in a high field behave as if their temperature were increased many times with respect to the temperature of the lattice. By applying a voltage pulse across the horizontal bar of a T-shaped sample, a Seebeck potential can be measured across the vertical bar, showing that the electrons are "heated" by the field in the horizontal bar.

Various kinds of related measurements are in progress in several laboratories, and we can hope that a semiempirical theory, in which all scattering mechanisms are carefully included, will be justified.

A bibliography on this subject of current interest, complete until 1958, will be found in the papers by S. H. König: "Hot and Warm Electrons—A Review," *J. Phys. Chem. Solids*, **8**, 227 (1959), and E. Conwell: "Lattice Mobility of Hot Carriers," *ibid.*, 234.

General Bibliography

Given below are the most useful books and review articles in which extensive original references may be found.

Books and Reports

A. H. Wilson, *Theory of Metals*, Cambridge University Press, Cambridge, 1936.
F. Seitz, *Modern Theory of Solids*, McGraw-Hill Book Company, New York, 1940.
L. Brillouin, *Wave Propagation in Periodic Structures*, McGraw-Hill Book Company, New York, 1946.
W. Shockley, *Electrons and Holes in Semiconductors*, Van Nostrand Company, New York, 1950.
M. Born and K. Huang, *Dynamical Theory of Crystal Lattices*, Oxford University Press, Oxford, 1954.
R. Peierls, *Quantum Theory of Solids*, Oxford University Press, Oxford, 1954.
C. Kittel, *Introduction to Solid State Physics* (2nd Ed.), John Wiley and Sons, New York, 1956.
J. C. Slater, *Electronic Structure of Solids Technical Reports*; *Solid-State and Molecular Theory Group*, M.I.T., Cambridge, Mass.
    4. *The Energy Band Method*, 1953.
    5. *The Perturbed Periodic Lattice*, 1953.
    6. *Configuration Interaction in Solids*, 1954.
    10. *The Interaction of Waves in Crystals*, 1957.
A. J. Dekker, *Solid State Physics*, Prentice-Hall Inc., New York, 1957.
W. Ehrenfeld, *Electrical Conduction in Semiconductors and Metals*, Oxford University Press, Oxford, 1958.

Review Articles

F. Herman, "The Electronic Energy Band Structure of Silicon and Germanium," *Proc. I.R.E.*, **43**, 1703 (1955).
"Proceedings of the International Conference on Electron Transport in Metals and Solids," *Can. J. Phys.*, **34** (1956). (Contains a dozen papers of fundamental importance.)
Encyclopedia of Physics, Springer-Verlag, Berlin.
    Vol. 1, *Crystal Physics I*, 1955. In particular:
        G. Liebfried, "Gittertheorie der mechanischen und thermischen Eigenschaften der Kristalle," p. 104.
        M. Blackman, "The Specific Heat of Solids," p. 325.

Vol. 19, *Electrical Conductivity I*, 1956. In particular:
J. C. Slater, "The Electronic Structure of Solids," p. 1.
H. Jones, "Theory of Electrical and Thermal Conductivity in Metals," p. 227.
Vol. 19, *Electrical Conductivity II*, 1957. In particular:
O. Madelung, "Halbleiter," p. 1.
Solid State Physics (F. Seitz and D. Turnbull, Editors), Academic Press, New York.
Vol. 1, 1955.
J. R. Reitz, "Methods of the One-Electron Theory of Solids," p. 2.
H. Y. Fan, "Valence Semiconductors, Germanium and Silicon," p. 284.
D. Pines, "Electron Interaction in Metals," p. 368.
Vol. 2, 1956.
J. de Launey, "The Theory of Specific Heats and Lattice Vibrations," p. 220.
Vol. 4, 1957.
F. J. Blatt, "Theory of Mobility of Electrons in Solids," p. 200.
T. O. Woodruff, "The Orthogonalized Plane Wave Method," p. 367.
Vol. 5, 1957.
W. Kohn, "Shallow Impurity States in Germanium and Silicon," p. 258.
Vol. 7, 1958.
P. G. Klemens, "Thermal Conductivity and Lattice Vibrational Modes," p. 1.

Index